Green Energy and Technology

For further volumes:
http://www.springer.com/series/8059

Lucia Recchia · Paolo Boncinelli · Enrico Cini
Marco Vieri · Francesco Garbati Pegna
Daniele Sarri

Multicriteria Analysis and LCA Techniques

With Applications to Agro-Engineering Problems

 Springer

Dr. Lucia Recchia
Dipartimento di Economia,
Ingegneria, Scienze e Tecnologie
  Agrarie e Forestali
Università degli Studi Firenze
Piazzale delle Cascine 15
50144 Florence, Italy
e-mail: lucia.recchia@unifi.it

Prof. Marco Vieri
Dipartimento di Economia,
Ingegneria, Scienze e Tecnologie
  Agrarie e Forestali
Università degli Studi Firenze
Piazzale delle Cascine 15
50144 Florence, Italy
e-mail: marco.vieri@unifi.it

Dr. Paolo Boncinelli
Dipartimento di Economia,
Ingegneria, Scienze e Tecnologie
  Agrarie e Forestali
Università degli Studi Firenze
Piazzale delle Cascine 15
50144 Florence, Italy
e-mail: paolo.boncinelli@unifi.it

Dr. Francesco Garbati Pegna
Dipartimento di Economia,
Ingegneria, Scienze e Tecnologie
  Agrarie e Forestali
Università degli Studi Firenze
Piazzale delle Cascine 15
50144 Florence, Italy
e-mail: francesco.garbati@unifi.it

Prof. Enrico Cini
Dipartimento di Economia,
Ingegneria, Scienze e Tecnologie
  Agrarie e Forestali
Università degli Studi Firenze
Piazzale delle Cascine 15
50144 Florence, Italy
e-mail: enrico.cini@unifi.it

Dr. Daniele Sarri
Dipartimento di Economia,
Ingegneria, Scienze e Tecnologie
  Agrarie e Forestali
Università degli Studi Firenze
Piazzale delle Cascine 15
50144 Florence, Italy
e-mail: daniele.sarri@unifi.it

ISSN 1865-3529
ISBN 978-1-4471-2709-3
DOI 10.1007/978-0-85729-704-4
Springer London Dordrecht Heidelberg New York

e-ISSN 1865-3537
ISBN 978-0-85729-704-4

British Library Cataloguing in Publication Data
A catalogue record for this book is available from the British Library

Cover design: eStudio Calamar, Berlin/Figueres

Printed on acid-free paper

Springer is part of Springer Science+Business Media (www.springer.com)

The book manuscript has been edited by Paolo Boncinelli and Lucia Recchia. All LCA computations presented in the text have been carried out by Lucia Recchia.

(Authors' Note)

Contents

Acronyms

ANPA	Italian National Agency for Environment Protection
ARSIA	Tuscany Regional Agency for Development and Innovation in the Agricultural sector
AU	Average Utilization
CER	Cumulated Energy Requirement
CPO	Crude Palm Oil
CTI	Italian Heat Technology Committee
EC	European Commission
EEA	European Environmental Agency
EFB	Empty Fruit Bunches
EIA	Environmental Impact Assessment
ENEA	Italian National Agency for New Technologies, Energy and Sustainable Economic Development
EU	European Union
FFB	Fresh Fruit Bunches
FTS	Full Tree System
GHG	Greenhouse Gases
GWP	Global Warning Potential
ISTAT	Italian National Institute of Statistics
ITABIA	Italian Biomass Association
kW_{th}	Thermal Kilowatt
LCA	Life Cycle Assessment
LHV	Lower Heating Value
MCA	Multi-Criteria Analysis
NDP	Nominal Device Power
PKO	Palm Kernel Oil
POME	Palm Oil Mill Effluent
PTO	Power Take Off
RED	Renewable Energy Directive

SFP	Specific Fuel Consumption
t_{db}	Tons Dry Basis
t_{wb}	Tons Wet Basis
UT	Usage Time
VOC	Volatile Organic Compounds

Chapter 1
Introduction

1.1 General Introduction: Purposes and Organization

In recent years, the development of both numerical algorithms and computer power has made it possible to extend the use of numerical methodologies to analyze and elaborate data in sectors different from traditional ones (mathematics, physics and industrial engineering). This is the case of agro-engineering processes, concerning both field and industrial operations.

In agriculture, the empirical approach is still widely used to solve problems and to optimize the different phases of the production chain. This traditional approach keeps preserving its effectiveness and popularity due to the extreme variability of scenarios available, as well as to the lack of consistent and homogeneous data for each phase of the chain. However, the use of numerical analysis is quickly gaining ground, progressively supporting and ousting traditional methods, so that, nowadays, a certain level of abuse in the exploitation of numerical methodologies is observed. The robustness and efficiency of numerical tools are often considered as a sufficient reasons for performing extensive computations to investigate different scenarios without any preventive selection based on a critical analysis of the problem. As an example, the problem of identifying a function by means of patterned data sampling can be mentioned. The general approach was to acquire an amount of data as large as possible. After the discovery of the Shannon theorem, the collection of a limited number of data was found to be sufficient to reconstruct all kinds of functions, whether periodical or otherwise. For this reason, a reflection on the correct application of numerical methodologies seems to be necessary, with the aim of investigating the essential key-points of problems, without losing oneself in considering minor and irrelevant details. As a consequence, research in the agro-engineering sector has lately concentrated its efforts on developing methodologies capable of performing analysis of specific aspects of agricultural processes, obtaining definite answers and, at the same time, allowing one to save time and resources by avoiding useless computations.

L. Recchia et al., *Multicriteria Analysis and LCA Techniques*,
Green Energy and Technology, DOI: 10.1007/978-0-85729-704-4_1,
© Springer-Verlag London Limited 2011

Nowadays, one of the most critical problems to be considered in all production processes is represented by the environmental impact of each phase of the working chain, in terms of both consumption of non-renewable resources and raw materials (such as fossil fuels), and emission of greenhouse gases, by-product reuse and waste disposal. This is also the case in the agro-food industry. Over the last decade, the role of agriculture and, consequently, farms has slowly changed. Originally, farms were only considered as technical-economic units exploiting the land resources to implement agricultural, forestry and/or zootechnical productions. At present, farms have acquired important tasks such as landscape conservation, land coverage and environmental protection against various types of pollution. These new tasks have been associated with farms not only by national and European policies but also by market needs. A typical example is provided by agriculture in Tuscany, Italy, whose high-quality products are associated with the culture of the specific territory in which they are produced, too. Thus, it is not unlikely, in the near future, to think of some kind of environmental certification characterizing these products as an added value to their quality.

In this framework, the authors of this book have spent significant efforts during last years to establish adequate methodologies aimed at evaluating the environmental sustainability of agro-engineering processes, in terms of both their energetic costs and environmental impacts. This is not an easy task to accomplish, since such processes are extremely heterogeneous, due to the variety of environments in which crops grow, to the different typologies of cultivars of each crop, to the various levels of mechanizations in fields, and so on. Such an approach is based on the logic of a total-quality philosophy applied to the production chain, where quality means optimization of resources required for performing process operations, providing at the same time useful information for strategic planning.

Life cycle assessment (LCA) methodology has proved to be one of the most effective tools for carrying out this kind of analysis, and, for this reason, it has become very popular. However, as discussed before, the wide variability and complexity of possible scenarios often determine a huge amount of configurations to be investigated, which require considerable computational time and resources, making it difficult to use LCA in practical applications. As a consequence, some sort of pre-filtering is required which should be capable of selecting the most relevant cases to be investigated by means of LCA.

Recently, some of the present authors have been involved in a study concerning optimal plant configuration for the management of riparian vegetation in Tuscany, in the area of Chianti in Florence Province, and its possible reuse as biofuel, evaluating the benefits and drawbacks from the economical, environmental and managerial points of view [1]. The lack of funds available for performing this study pushed the authors to develop and tune an innovative approach based on the joint use of multicriteria analysis (MCA) and LCA. This approach is the subject of this book.

In the previously cited study, the application of MCA for the analysis of possible chain scenarios allowed one to select the main targets in terms of energetic requirements, dramatically reducing the global number of chain configurations to be investigated by means of LCA. Following this approach, it was also possible to increase the number and typology of scenarios under investigation, extending the sphere of the analysis and completing it with additional results which would have been impossible to achieve otherwise. Results show how the proposed methodology, coupling MCA with LCA, is robust and, at the same time, easy to implement, and provides one with a clear view of the most suitable solutions. Moreover, it allows the operator to establish a precise, reliable and repeatable procedure for evaluating different scenarios in a single way. In the authors' opinion, this represents the most valuable merit of the proposed methodology.

As its main purpose, this book would like to represent a useful reference for all engineers, researchers and high-level students who might be interested in applying LCA as an effective tool for their professional and/or academic activities in the agro-engineering field. In particular, an appropriate and systematic use of LCA methodology could be of interest for those farmers who are determined to follow a different approach in the working chain of their products, pursuing a target of "total quality". Actually, this requirement has become more and more essential for being competitive on global markets, which more and more often require a certification of quality taking into account energy consumptions and environmental impacts associated with each product. On the agro-industrial side, too, LCA could be useful to managers for establishing and controlling process optimization. Keeping in mind these basic concepts, which represent one of the most important research activities of the group, especially for Tuscany farms, the authors have produced this book, which provides a wide-ranging view on the application of both MCA and LCA to different contexts in agriculture.

The book contents are organized as follows. In Chap. 2, the general theory of both MCA and LCA techniques is explained and discussed. In the following chapters, some applications of the proposed methodology are presented, in order to provide the reader with some practical examples: energetic use of biomass and biofuels (Chap. 3); agricultural and forestry mechanization (Chap. 4); olive oil production chain (Chap. 5); oil palm production chain (Chap. 6). Such applications are discussed in detail with the aim of improving and deepening the reader's knowledge on the subject, guiding her/him in implementing and applying the methodology to her/his field of interest.

Reference

1. Recchia L, Cini E, Corsi S (2010) Multicriteria analysis to evaluate the energetic reuse of riparian vegetation. Appl Energy 87:310–319

Chapter 2
General Theory of Multicriteria Analysis and Life Cycle Assessment

2.1 Objectives of the Proposed Methodology and Its Application

As reported in the Chap. 1, one of the most critical problems to be considered in all production processes of the agro-industrial sector is represented by the environmental impact of each phase of the working chain, in terms of both consumption of non-renewable resources, emissions of greenhouse gases, by-product reuse and waste disposal. Moreover, these environmental aspects must be evaluated assuring the technical feasibility and the economical sustainability of the proposed solutions.

This is not an easy task to accomplish, since such processes are extremely heterogeneous, due to the variety of environments in which crops grow, to the different typologies of cultivars of each crop, to the various levels of mechanisations in fields, and so on. The life cycle assessment (LCA) methodology has proven to be one of the most effective tools for carrying out the environmental analysis, even if the large variability and complexity of possible scenarios often determine a huge amount of configurations to be investigated, which require considerable computational time and resources. Therefore, some sort of pre-filtering is required which should be capable of selecting the most relevant cases to be investigated by means of LCA.

As a consequence of the previous considerations, the innovative approach proposed in this book is based on the implementation of the multicriteria analysis (MCA) and the LCA: particularly, the application of the MCA to the alternative solutions allowed to select the most suitable ones in terms of economical and environmental sustainability, dramatically reducing the global number of chain configurations to be investigated by means of LCA.

In the following paragraphs the fundamentals of the two methodologies are briefly illustrated, reporting also some indication about the main common choices adopted in the development of the proposed applications.

L. Recchia et al., *Multicriteria Analysis and LCA Techniques*,
Green Energy and Technology, DOI: 10.1007/978-0-85729-704-4_2,
© Springer-Verlag London Limited 2011

2.2 Generals About the MCA

The development of the MCA is very recent and has been carried out during the last three decades with the aim to consider several consequences of proposed solutions of various typologies of problems. Particularly, the MCA has been introduced after having relieved that intuitive solutions are often not the most suitable and even if they are profitable for a specific aspect they could not be for another one. In fact, the MCA has been introduced because of the necessity to develop multiple evaluations at the same time, taking into account different points of view highlighted by different typologies of stakeholders. Therefore, this methodology can be classified as a supporting tool for decision makers because it is not able to identify the right solution whilst it is useful to organise all the available information, to supply a possible interpretation and to check the pros and cons associated with all the alternatives.

The decision process constitutes several steps: at first the different options must be identified; then a group of parameters to be used to compare the alternatives must be set; finally, all the scenarios must be judged regarding the fixed criteria with the aim to identify the most suitable options.

It is obvious that this approach may be applied in several sectors and whenever it is necessary to carry out a choice in a decision process. However, the methodology can be developed with a different type of detail according to the stage and complexity of the decision process. In fact, if the decision process is carried out at a planning stage, data available for each hypothesised scenario present lower quality and quantity than those which can be collected at successive design stages (i.e. feasibility study). Moreover, the specific sector where the decision process is developed implies a different level of complexity and a different set of criteria which can be applied more or less easily.

The MCA is not uniquely defined and a lot of techniques have been developed with the aim to better adapt the methodology to the specific problem to be solved, including all the preferences promoted by different stakeholders.

Anyway, the main structure of the MCA provides to set all the possible alternatives and to define the criteria to be used for the evaluation.

Particularly, the MCA has a precise structure that includes several steps (see Fig. 2.1):

1. Problem identification and objectives definition;
2. Problem structuring, defining both the options and the criteria to be used;
3. Preference modelling, where scoring and weighting are carried out;
4. Aggregation and analysis of the results;
5. Discussion and negotiation about the obtained results.

The first step is the identification of the problem under discussion, defining also the goals and scopes of the analysis. In this phase it is important to consider all the laws, local constraints and policies that usually highlight the most critical aspects and cause the most important differences between the hypothesised alternatives.

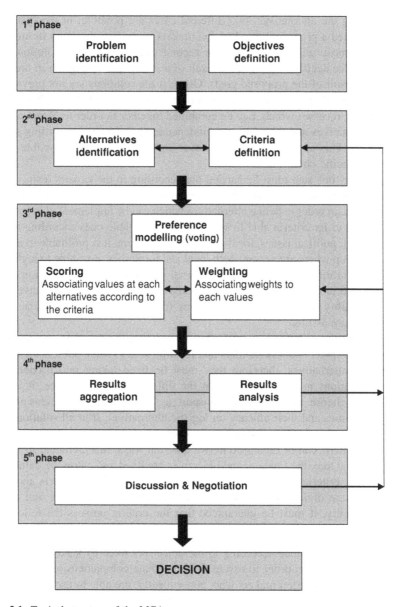

Fig. 2.1 Typical structure of the MCA

In fact, right decisions can be made only if the objectives to be achieved are clearly defined. Therefore, the goals must be specific and measurable, although they could be time-dependent, i.e. reachable in the brief, medium or long period.

Afterwards, the problem is structured fixing both the possible alternatives and the criteria to be used to evaluate and compare them.

Concerning the alternatives, two different cases are possible: in the first case they are decided a priori and the decision makers must compare them in order to indicate the most suitable ones; in the second case the possible solutions are identified by the decision makers as a result of a systematic discussion in order to assure the pursuit of the proposed goals. Often, if the solutions are not previously defined, a rational methodology may be implemented dividing the analysed process into sub-processes which may be combined together in order to obtain all the possible alternatives and which may be independently evaluated according to the fixed criteria. This approach assures to cover and to assess all the possible solutions for a specific chain.

In any case, this step must be carried out according to the experts team which must have sufficient knowledge of the problem to be solved and of the site characteristics, in order to define alternatives which can be implemented in specific situations and to fix criteria able to select the most suitable ones according to the technical level, political issues, local needs, etc. Therefore, it is profitable to assure a working group where are present both local and foreign experts, able to highlight the site peculiarities and to supply an external perspective of the situation. Reference [2] also indicates the benefits originated by teams where gender diversity, mixed nationality and different perspectives (e.g. politicians, technicians, academics, etc.) are assured.

Concerning the criteria, it must be considered that they explain the point of view of both experts and stakeholders and must be able to carry out a comparison between the alternatives, therefore they must be fixed taking into account the proposed solutions in order to highlight the differences. For instance, it is not profitable to choose the means of transport as an indicator for assessing the transport facilities and their efficacy on logistic alternatives, if in all solutions the same mean or very similar typologies are adopted.

Usually, criteria can be organised in two different ways. In a hieratical structure known as value tree, the fundamental objectives are fixed and cause the definition of the specific criteria. Alternatively, it is possible to list all the criteria and in a successive phase divide them into groups characterised by the same aim to be pursued. Anyway, it must be guaranteed that the criteria possess the following properties:

- Certain value relevance according to the objectives fixed by stakeholders;
- Understandability in order to assure the immediate comprehension of the criteria by all the decision makers who, consequently, are able to use them in the evaluation and comparison of processes regarding each proposed solution;
- Measurability, meaning that at least one indicator measurable in a qualitative or quantitative way corresponds to each criterion, otherwise the criterion must be considered non usable;
- Completeness because the criteria all together must be able to cover all the proposed aims and must be able to evidence all the possible differences. This scope could not be reached at the first stage and some lack was evident in the indicators' definition only at the final stage when the different solutions are confronted.

Fig. 2.2 List of criteria
characteristics and indication
of subjects involved in their
definition

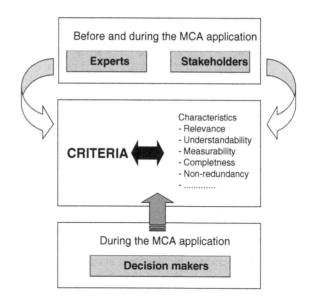

Particularly, the comparison allows verifying if the results agree with the expected
outputs; if some incoherent result occurs a specific analysis must be done and
eventually it may be interesting to introduce other indicators, which may better
respond to the objectives. For this reason also the decision makers may contribute
to modify or integrate the criteria, but only in a second phase basing on the
assumptions of experts and stakeholders (see Fig. 2.2);

- Non-redundancy avoiding that two different criteria using indirectly the same
  indicator overestimating a specific aspect. Moreover, it is also important to
  check if there are some unnecessary criteria, in order to improve the economical sustainability and to reduce the time needed to develop the MCA.
  However, it is possible that the double counting is done because the same
  indicator implies several effects from different points of view: for instance, the
  transport distance expressed in kilometres may be a useful indicator both for
  environmental and economical aspects; therefore, in this case considering twice
  the benefits due by reduced transport distance could be considered correct and
  may clearly highlight the profitability associated with solutions that adopt this
  logistic approach.

In addition, it must be suggested to use a limited number of criteria. Firstly,
problems associated with the correct understandability of certain criteria may be
avoided by the stakeholders excluding those criteria, that may result as too much
technical and may require a very high level of knowledge of the environmental
and/or economical issues. On the other hand, a reduced size of criteria may help to
limit the risks of redundancy and double counting, contributing to permit a more
easy comprehension of the expected effects of certain decisions. Finally, in this
way it is possible to guarantee a better communication of the obtained outputs of
the MCA.

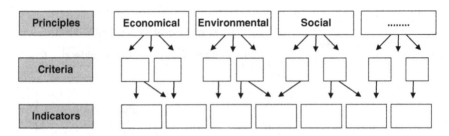

Fig. 2.3 Description of the phase of the criteria identification in the MCA

In general, the choice of the criteria and the relative indicators must be carried out considering that firstly the principles or objectives must be fixed, then the typology of the criteria must be set and finally the quantitative or qualitative indicators must be identified. This process is summarised in Fig. 2.3; moreover, some parameters may be identified in order to validate the values of the indicators defined for the evaluation of the alternatives. For instance if the environmental sustainability must be evaluated for a specific process, an adopted criterion can be the minimisation of the GHG emissions and the associated indicator can be the amount of fossil energy required: on the basis of a literature review a range of variation of this indicator must be fixed in order to facilitate the validation of the values obtained for all the alternatives and to set the suitability classes for the judgement phase.

Particularly, the principles are determined by the needs of the specific sector where the options are developed: for example, if the problem is to compare different techniques of olive oil milling, it is important both to evaluate the sustainability of the production and to assure adequate product performances according to the laws and market requirements. The subsequent step requires the definition of the criteria used to verify the sustainability of the olive oil chain: considering the most critical aspects of the agro-industrial sector, it is possible to focus the analysis on economical and environmental issues: the first ones must guarantee the convenience of the product in comparison with other products referred to in the same trade segment; whilst the second ones must highlight some benefits of the product in order to preserve the environment, making the oil more appreciable to the customers and allowing emissions within the legal thresholds. In fact, the principles become operative through the criteria, even if the criteria are not able to produce a direct measurement of the suitability of a solution as the indicators can do: the indicators are variables or parameters associated with each criteria, which allow evaluating easily the alternatives in a qualitative or quantitative way. For instance, for the olive oil production an environmental criteria could be the pressures on the global warming phenomenon and a quantitative indicator is obviously the CO_2 equivalent emissions. However, taking into account the complexity of the calculation of this indicator, it is possible to promote the use of other indicators, that may also be considered reliable on the basis of literature and experimental tests: for example, it is possible to choose as alternative indicators the fertilisers requirement in the olive-grove and the electricity consumption in the milling.

Finally, it is important to verify the values of the indicators identifying the range where they can vary. This approach may help to improve the comprehension of the indicators themselves. Moreover, a specific identification of certain monitoring procedures on the indicators may also be introduced: if the water abstraction is an indicator used in order to evaluate the environmental sustainability, it may be interesting to provide an analysis of the periodic variation of the local water availability.

Particularly, the set of criteria must be done by experts according to the decision makers: the criteria must be chosen considering the most significant aspects in relation with the priorities for optimising the choice. For instance, if design alternatives are compared, aspects concerning economics, efficiency, reliability, environment or other aspects could be analysed, but a context analysis occurs to select the most important ones. This analysis must be developed taking into account the expected benefits and the barriers, that may contrast the realisation of the proposed options.

In addition, it must be highlighted that the nature of the criteria is heterogeneous: for example, it is possible that a specific set of criteria concerns environmental impacts, another economical aspects and another health risk. When for each criterion a relative set of indicators has been fixed, it is possible to obtain a judgement of each option for each precise aspect investigated. However, the objective of the methodology is to summarise in a global judgement the evaluation of each alternative, in other words it is necessary to define relative weights able to consider the relative importance of the various aspects. This is a very difficult phase where the decision makers must justify why a specific set of criteria must be considered more important than the others. Usually, this activity is supported by national and international laws and regulations, and is conducted according to the planning development and local needs: for instance if European and national policies have indicated as a priority the reduction of GHG emissions, probably the environmental criteria should result as more important than the economical ones, which evaluate the local convenience of the alternatives, e.g. the pay-back period of the investment needed for their implementation. Moreover, the criteria should be chosen in such a way that they are able to quantify the different impact of the options regarding a particular aspect: if different typologies of crop production are analyzed and the environmental pressures must take into account the associated GHG emissions, the amount of diesel fuel used during field operation, the fertilisers requirement and the quantity of the pesticides used may constitute accurate indications. On the other hand, these criteria may also supply additional information about other environmental aspects: higher amounts of fertilisers imply higher risks in terms of nutrient leaching and eutrophication, while the reduced use of pesticides limits the biodiversity losses.

Besides, some relative weights can also be established within a set of criteria in order to indicate which indicator is considered the most significant to evaluate a specific aspect of the examined options (see Fig. 2.4).

Fig. 2.4 Description of
different typologies of
weighting of the proposed
criteria

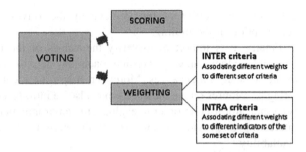

Particularly, an MCA and the weighting phase may be developed by adopting a
ranking or a rating methodology. The ranking methodology associates with each
decision element a corresponding degree of importance, for example adopting
specific questionnaires to be directly filled in by the decision makers. On the other
hand the rating methodology provides to assign to each decision element a corresponding score (e.g. from 0 to 1 or from A to H) obtained evaluating all the
identified criteria and summarising them in a global and unique score.

Moreover, it is important to highlight that each criteria can be applied using
quantitative or qualitative indicators; an MCA can include both these typologies of
indicators at the same time. This characteristic is very useful in a previous stage of
the study when data of different levels of accuracy can be available to describe
different aspects of the alternatives proposed.

Considering that each application of a specific indicator on the hypothesised
alternatives implies the assignment of a relative score, alphabetical or numerical,
it is important to know the range of variation of each indicator. Particularly, this
knowledge allows to classify the obtained score with the aim to identify easily the
suitability level of the examined option. The classes may be characterised by
values of different nature: also in this case the values may be quantitative or
qualitative, alphabetical or numerical. Usually, with the aim to promote an easy
and immediate comprehension of the results, the numbers of classes are limited:
for instance, if three classes are fixed it is possible to use high–medium–low or
A–B–C or 1–2–3. Fig. 2.5 shows that from each typology of criteria the useful
indicators have been identified in order to relieve the options characteristics; once
the scoring of the indicator has been done, the associated value permits the
identification of the relative class.

Therefore, the range where these indicators may vary must be previously
known in order to define correctly the sustainability classes: in other words, this
phase hypothesises a good knowledge of the sector where the alternatives are
defined.

In any case the possibility to adopt a classification method for the obtained
indicators allows to compare several typologies: once specific intervals of variation for each indicator have been established to reach the proposed objectives, the
same sustainability class may be attributed to qualitative or quantitative indicators
affecting different aspects (i.e. environment, economics, etc.).

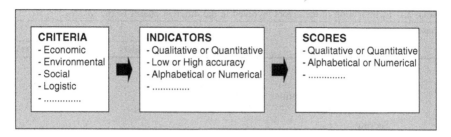

Fig. 2.5 Definition of criteria, indicators and scores evaluating the sustainability

Considering all the previous assumptions, the iterative nature of the MCA is quite evident in order to guarantee the mentioned properties of the used criteria and to improve the quality level of the study: for instance, with the aim to assure the completeness of the study, the introduction of a new criterion may be needed, otherwise to avoid the redundancy the merging of criteria able to describe similar effects may occur. Moreover, if the application of the MCA starts at the planning stage it must be considered that solutions identified as the most suitable at the beginning have to be monitored for several subsequent years in order to evaluate the effects produced: in this case the data of the MCA are integrated during the years, while the objectives fixed at the planning stage are verified in the successive steps (e.g. feasibility study) and eventually modified considering all the new inputs available.

Finally, time could be an important aspect in order to define the criteria of the MCA: for instance, some indicators could be described through a temporal function and assume different values or scores at the time passing. Particularly, some alternatives may obtain good scores at different time than others and, for example, may be profitable in the brief, medium or long period. Therefore, if the indicators require a specific time interval where they have to be observed, the time horizon used for the application of the MCA must be set in order to decide a priori, the most suitable period for comparing all the different options. In fact, it is important to assess both temporal and permanent effects associated with the alternatives, taking into account that different criteria may have different time horizons: economical criteria usually have a brief-medium horizon (e.g. the evaluation through the pay-back period in a maximum time of 5 years), whilst environmental ones may require a long term horizon (e.g. the carbon stock variations require a period of 20 years).

The next phase is about preference modelling: the decision makers provide to apply the criteria to the proposed alternatives and to calculate the relative indicators. This process may induce several problems because during its development it is possible to highlight some critical aspects of the previous stages: for example, the decision makers may indicate additional solutions or criteria, which may allow a better development of the MCA in order to reach the proposed objectives. All the assumptions originated during this phase may be integrated in the MCA according

Table 2.1 Identification of the actors involved in the different phases of the MCA

Phase	Activity	Actors		
		Experts	Stakeholders	Decision makers
1	Problem identification		X	
	Objectives definition	X	X	
2	Alternatives identification	X		X
	Criteria definition	X	X	X
3	Scoring			X
	Weighting			X
4	Result aggregation			X
	Result analysis	X	X	
5	Discussion and Negotiation		X	X

to its iterative nature, if considered necessary both by the experts and the stakeholders.

With the aim to avoid any misunderstanding it is necessary to highlight that the decision makers are the persons, who have the actual responsibility of the final decision and must indicate the most suitable solutions for the considered problem. They are required to take into account the opinions of both experts and stakeholders, but they cannot be catalogued as experts or stakeholders. On the other hand, the experts may be identified as technicians or academics working in the specific expertise sector of the problem under investigation, whilst the stakeholders are any single individual or group of people, who have an interest (e.g. a social or economical interest) in the examined problem. Table 2.1 identifies the actors involved in the different phases of the MCA.

It is important that, before the voting takes place, each decision maker has the possibility to expose to the others his opinion about the criteria and the relative indicators: particularly, it may be useful to know the indicators that are considered the most significant ones and why it may be affirmed this basing on the different points of view of the decision makers. Once this discussion has been carried out, each member must individually associate the scores with the various solutions, without declaring their choices in order to avoid any influence between the members themselves. The pre-voting discussion is fundamental because it is able to guarantee the interdiscipline of the process, promoting at the same time

- Compromises to accommodate the different needs;
- Interest in knowing all the different points of view;
- Respect for the competence of the other team members;
- Agreement between members about the objectives of the work.

Moreover, in this phase it is possible to decide to use some weights for the different typologies of the adopted indicators: these weights may be assigned a priori by experts on the basis of the literature or policies indications, or alternatively they may be indicated by each decision maker during the judgements definitions. In fact, weights may constitute a subjective approach to the problem

Fig. 2.6 Different approach
for scoring phase

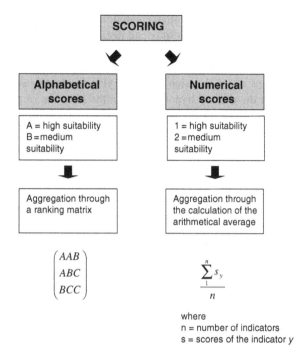

and may be very different from one decision maker to another according to their
background (e.g. technicians vs. politicians). For this reason, in some cases, the
weights are previously decided as a compromise between all the different points of
view, taking into account the relative importance of the indicators on the basis of
the quality of the available data or of the importance of the associated criterion in
respect to the other ones.

Once the judgements have been defined for each analyzed aspect of the proposed solutions, the aggregation of all the indicators scores must be done, taking
into account also the associated weight. For each alternative the combination of all
the assigned scores may be carried out following specific rules, that may differ
very much because of the nature of the scores: particularly, if the alphabetical
scores have been applied, the use a specific ranking matrix is needed, whilst,
if numerical scores have been fixed, the associated arithmetic average can be
calculated (see Fig. 2.6). Particularly, if the MCA is carried out using a numerical
approach to identify the most suitable alternatives proposed to solve the problem,
at the first stage scores are associated with each solution (scoring), whilst at the
second stage some weights are also applied (weighting).

It must be highlighted that the application of specific weights is possible only if
the independence between used criteria is assured. However, in this case, the MCA
is a compensatory technique, because high values obtained on one criterion are
compensated by low values on another.

When the most suitable solutions have been identified, stakeholders must analyze the obtained results and decide if there are some mistakes or unexpected outputs and if the solutions may be considered as satisfying. Obviously, if some problems or ambiguities are relieved, it is possible and necessary to reconsider all the steps of the decision process trying to identify some possible misunderstandings in the defined hypothesis or in the structure of the analysis.

Finally, the results must be discussed and it might be necessary to propose a negotiation between the stakeholders in order to reach a common interpretation of the results with the aim that all the stakeholders agree about the most suitable solutions identified by the MCA.

2.3 Generals About the LCA

The development of the LCA methodology began in the 1960s, with the aim to evaluate the problems related to raw materials and energy supplies in the industrial sector. One of the first studies developed following the LCA methodology was carried out for the Coca-Cola Company to compare the environmental benefits and drawbacks associated with the use of plastic or glass bottles [4]. This study quantified the raw materials and fuels needed and the environmental pressures due to the manufacturing processes for each beverage container. Afterwards other companies in both the United States and Europe performed similar comparative life cycle inventory analysis.

The process of quantifying the resource use and environmental releases of products had started to be known in the United States as a 'resource and environmental profile analysis' (REPA) while in Europe it was called an 'ecobalance'.

Currently, the LCA interests a large number of companies and industrial sectors, that use this approach to choose between alternative solutions during the design phase and/or the production phase. Also some governments and institutions have started to adopt the LCA as a valid instrument to associate with each product or service the environmental impacts caused during its production: the recent RED indicates the LCA to evaluate the benefits obtained through the use of biofuels instead of traditional fossil sources; the Ecolabel certifications are based on this approach; the Carbon Footprint methodology is directly linked and derived by the LCA, etc. In fact, during the previous decades customers and consequently the market has started to require additional information about the sustainability of the production processes, even if various researches demonstrate that all these data supplied by the industries do not significantly influence the trade and the relative profits. Moreover, because of the inappropriate use of the LCA by manufacturing industries, that begin to associate marketing claims with their products highlighting partial environmental benefits (the so—called "greenwashing" phenomenon), a standardisation process of the methodology was started. During the SETAC congress in 1993, the LCA name and acronym was fixed according to the first definition of the methodology: life cycle analysis is an objective instrument able to evaluate the energetic and environmental loads for a process or activity, carried out through the definition of materials, energy and wastes flows. In addition, it is obvious that the assessment includes the whole life cycle of the process or activity, starting from the extraction and treatment of the raw materials, up to the manufacturing, transport, distribution, use and reuse, recycle and the final disposal. Finally, the procedures of the LCA were standardised in 1997 by the International Standard Organisation (ISO) through the ISO 14040 series updated in 2006.

The LCA is one of the tools of environmental systems analysis and it "provides a systematic framework, that helps to identify, quantify, interpret and evaluate the environmental impacts of a product, function or service in an orderly way" [11]. This technique allows to quantify the total environmental impacts of the provision of a product or service from original resources to final disposal, or so-called "cradle-to-grave". LCA is mainly a tool used for describing environmental impacts. Examples of other environmental systems analysis tools include risk assessment, environmental impact assessment (EIA), material flow analysis, environmental auditing, etc., but the LCA results as unique for its "cradle-to-grave" approach combined with its focus on products, or rather the functions that products provide.

All inputs from and outputs to the natural system, such as resource extraction and emissions, must be taken into account.

It is important to highlight some fundamentals of the LCA:

- This methodology is not able to analyze all the environmental pressures that a system can origin (e.g. landscape modifications, etc.), therefore usually the LCA results are integrated with evaluations obtained through the application of other methodologies (i.e. Environmental Impact Assessment);
- The LCA is a quantitative methodology, i.e. the impacts estimation is structured adopting specific indicators able to quantify the associated environmental impacts through numeric values;
- The LCA is used to evaluate the environmental impacts from a global point of view. In fact, this methodology considers all the processes that have permitted to obtain a specific product, and also all the processes concerning the extraction, treatment and by-products disposal of the raw materials needed. Therefore, no limits from a geographical or time point of view are previously fixed: for instance, the impacts due to the extraction process of a metal in Africa or South America are computed in the total impacts associated with a specific product which is produced in Europe during the final steps of the industrial chain;
- The LCA is a relative tool intended for comparison and not absolute evaluation, thereby helping decision makers compare all major environmental impacts when choosing between several alternatives.

For all these reasons the LCA is used to compare improvement options, to design a new product or to choose between comparable products. In fact one of the advantages of LCA is that it avoids "problem shifting" from one stage in the life cycle to another, from one environmental issue to another and from one location to another [12], because it takes the entire life cycle of the product and all extractions from and emissions to the environment during that life cycle into account. According to these assumptions the LCA often shows unexpected and nonintuitive results: for instance, roses cultivated in Kenya and delivered to London by airplane transport may present lower GHG emissions than others cultivated in greenhouses in Netherlands (see the carbon footprint calculated at the Cranfield University, UK, as reported in the Economist the 17 of January 2008). Therefore, it is obvious that it is not correct to consider only one impact to determine, which alternative is the most profitable: for the roses case, for example also water management, land use modification and biodiversity losses should be considered and results obtained for these additional indicators could completely modify the final choice.

With the aim to identify some pros and cons of the LCA, it is important to know that the methodology is structured in four main steps (see Fig. 2.7):

1. Goal and scope definition. During this phase the objectives of the study, the functional unit, the system boundaries, the data needed, the assumptions and the limits must be defined. Particularly, the functional unit is the reference unit used to normalise all the inputs and outputs in order to compare them with each other;
2. Inventory data. This step concerns the analysis of the material and energy flows and the study of the system working. On the other hand the data collection for the entire life cycle implies the modelization of the analysed system. Moreover, one of the most critical aspects of this phase is the quality of the inputs, which must be verified and validated in order to guarantee the data reliability and correct use;
3. Impact assessment, evaluating the potential environmental impacts associated with identified inputs and releases through specific indicators usually fixed at the international level;
4. Impacts assessment and interpretation. In this phase the analyst aims to analyse the results and discuss them, helping decision makers to take a more informed decision. In addition, this step may highlight some problems in the LCA development which needs a more detailed approach: for instance, it can be decided to improve the quality level of some data collected from the literature because they describe a process which significantly influences an environmental pressure and therefore a more elevated accuracy of them may guarantee less variability in the results. This mechanism of the LCA assures the improvement of the results in an iterative way.

Concerning the inventory data it must be highlighted that the level of detail of the data collected determines the accuracy of the LCA results. However, the level of detail required to create the inventory depends on the size of the system and the purpose of the study: in a large system involving several industries with different production processes, certain details may not be significant contributors and may be omitted without affecting the accuracy or reliability of the results. These evaluations must be done during the goal and scope definition phase, taking into account the purpose of the work, the expected availability of the data, the financial resources and also the time needed to develop the study. In addition, the quality level of the inputs may significantly vary according to the specific objective of the LCA: if the LCA is developed at a feasibility level in order to compare different alternative scenarios, the data would be simply estimated or obtained from the literature; whilst if the LCA is applied to an industrial process well known and defined, the data could be directly measured or collected through questionnaires.

In any case, even if the uncertainty in the final results do not allow to establish if one proposed solution is better than another, it must be assumed that the most important scope of the LCA is providing the decision makers (e.g. government officials, multinational corporations, nongovernmental entities (NGOs), or, ideally, multi-stakeholder panels) with a better understanding of the environmental and health impacts associated with each alternative. However, these indications about the environmental impacts should constitute only one component in the decision process because the evaluation done by the LCA is partial and is not able to describe all the environmental aspects associated with a specific process nor furthermore to assess its economical and social implications.

Therefore, the LCA presents some limitations which must be considered in order to correctly evaluate the role of this methodology as a decision support tool.

Data uncertainty and use of database are not completely transparent. In fact in order to develop an LCA it is important to collect and choose all the data inputs available for the process to be modelled, but all the data associated with the sub-processes are usually assumed by existent databases. These databases may be associated with a specific LCA software or also available free [8] as open-source but, in any case, it is impossible for a user to verify all the sources and all the data needed: actually, it is obvious that a lot of data inserted in the LCA are out of the control of the analysts and for this reason it is always important to indicate the software and references used in the LCA.

The LCA is not uniquely defined from the methodological point of view and different options are available according to the ISO standard. In fact different possibilities exist to deal with the co-products and/or by-products that are produced along the life cycle of a product: the allocation and the credit method.

The allocation is based on the concept that environmental impacts caused by a process are caused by the co-products and the main product. Therefore, the impacts (e.g. emissions) are allocated proportionately between the co-products and the main product. The allocation can be based on various criteria such as mass, energy content (e.g. lower heating value) or market prices.

However, the ISO standards recommend to avoid allocation wherever possible by expanding the product system, i.e. to use the so-called substitution method. It takes into account the further utilisation of co-products which usually substitute conventionally produced goods of the same use. For instance, olive tree pruning adequately treated may replace the methane to produce thermal energy in a medium-sized boiler or may be landfilled allowing some savings in chemical fertilisers. In any case, since the conventional product does not need to be produced for these expected uses, the avoided environmental impacts associated with their production are credited to the main product.

Both approaches are characterised by pros and cons and can lead to considerably different results. The substitution methodology delivers a more exact and close-to-reality picture. However, calculation is relatively complex and may result in a great bandwidth of results depending on the system boundaries chosen. Moreover, the subtraction of the credits that are given in the substitution method, under certain circumstances can compensate all expenditures. The reason is that many different use (and thus substitution) options are possible for the co-products. On the other hand, allocation constrains this variability at least to a certain extent as the further use of the co-products is not taken into account. It thus typically delivers more transparent and unambiguous results. The fact that both methods may deliver significantly different results is due to the mathematical way of obtaining the results.

In addition, it must be considered that the substitution method is also called the expansion system method, because actually the system boundaries are enlarged to include also processes that describe the treatment and reuse of the co-products. Obviously the definition of these processes often is only hypothesised without any assurance regarding their effective working, therefore the relative data are, usually, simply estimated or collected from the literature.

As highlighted in [10] the identification of the system boundaries have to be made carefully also regarding energy and materials embodied in infrastructures and equipments used during the entire life cycle. The cut-off of the system modelization must be done using the literature data and also applying the LCA in an iterative way: particularly information collected about similar studies and a preliminary development of the LCA where values are used from an existing database without inserting more detailed data, can highlight that some inputs are not significant in the calculation of certain impacts. This is the case of the construction materials of an airplane when the scope of the study is to evaluate the CO_2 equivalent emissions: the emissions related to the flights will be as high so that all the emissions due to the production phase of the airplane could be correctly considered negligible.

Finally, the application of the LCA to the agro-forestry sector presents several criticisms.

Usually, field operations may be responsible for the major part of the environmental pressures if compared with transport phase or subsequent treatments of the agricultural products. This is the case of the SRF producing the wood chips (see Chap. 4) or of the olive oil milling (see Chap. 5), where the tillage operations cause more than 45% of the total amount of CO_2 equivalent emissions. Obviously it depends on the impacts that are taken into account, but generally the GHG confirms this trend together with other possible indicators as the primary energy requirements; in contrast other indicators (i.e. wastes production) may imply different results. In any case, considering these assumptions, all the inputs associated with the field phase highly influence the results, therefore high accuracy in collecting these data is required, even if it is not so easy because usually they present a large variability [1]: for instance the amount of fertilisers applied depends on soil fertility and typology, whilst the diesel fuel consumption is determined by the level of mechanisation, the power of the machines used, the morphological characteristics of the site, etc. In addition, although a lot of these data could be collected from the literature, it must be considered that all the references illustrate experimental data strictly referring to particular site conditions, which could be very different from those hypothesised for the LCA scenarios.

Additional criticisms may be introduced if some land use change must be detected. The LCA begins to include this analysis in its methodology when the production of bioenergy crops start to be studied in order to evaluate their sustainability. Particularly, determining the reference and the alternative land uses is fundamental in order to estimate some induced pressures on the environment, as carbon stock variations in soil and or vegetation. Moreover, since agricultural land is becoming increasingly scarce, more and more forest land or grassland is transformed into arable land. Such land use and land cover changes influence the area's carbon stock, i.e. the carbon content of both soil and vegetation. Any difference in carbon stock before and after cultivation has to be reflected in the greenhouse gas balances. Land use changes not only influence the climate but also factors such as biodiversity and habitat quality, soil functions and the water balance of a region. However, these impacts were not reflected by LCA up to now due to the lack of methodologies.

2.4   Generals About the Proposed Methodology

Considering all the MCA and LCA fundamentals, an approach able to integrate these two different methodologies has been developed. Particularly, at first the MCA has been applied in order to identify the more suitable solutions which in a second phase have been analyzed through the LCA in order to evaluate the environmental pressures more deeply.

In fact, for the environmental evaluation, it is needed to distinguish between local and global pressures that the agro-energetic chains may originate: for the first ones it is necessary to take into account the principles on which the EIA is typically based and which are included in the MCA; for the second ones the LCA methodology is needed.

In this way the effort and time needed for the LCA implementation have been significantly reduced because the methodology can be applied to a reduced number of scenarios once a large part of them has been classified as less profitable by the MCA.

The proposed applications illustrated in the following chapters regard very different chains of the agro-forestry sector. However, for all of them it has been hypothesised that they have been developed at the beginning of the planning stage or during the first step of the feasibility study. In this way the MCA and the successive LCA implementations have been used as a decision supporting tool in order to identify the most suitable design solutions.

Particularly, the MCA has been implemented considering the following steps:

- Goals definitions and decision makers typology;
- Scenarios definitions;
- Decision criteria identification;
- Criteria weights definition;
- Preference modelling obtaining the MCA results;
- MCA results interpretation.

During the goal definitions the operative scenarios have been defined considering the state of the art of the production or agro-energetic chains: in this phase, in order to evaluate all the different options and their different implications in terms of technical feasibility, cost/effectiveness analysis and environmental pressures, some literature review and experimental data collection have been carried out. In some cases the operative scenarios have been chosen hypothesising a specific geographical area, whilst in other cases the entire study has been based on generic data without considering the peculiarities due to the site specific characteristics. For instance, the yard scenarios in Chap. 3 have been based on the data of field operations and pruning management mainly referred to the Central Italy, whilst the Chap. 6 evaluates the convenience of establishing an oil palm plantation in an area where site conditions are not ideal without referring to a specific area around the world.

Taking into account that for each case study it has been hypothesised an application of the proposed methodology at the beginning of the design phase, the decision makers might be politicians, technicians and/or farmers.

All the applications have been structured first by identifying the main operative phases of each production or agro-energetic chain, secondly by defining for each phase all the possible scenarios and thirdly by combining together all the possible scenarios in order to define the whole production or agro-energetic chain, considering that each single choice in a specific phase may affect all the next ones. For each case study three categories of scenarios have been identified: one describing the field operations with particular attention to mechanisation, diesel fuel consumption and eventual use of nutrients and chemical compounds; one illustrating logistics and focusing on different means of transports, distances needed and storages organisation; one defining the plant of energy conversion or production.

It must be highlighted that in Chap. 6 the author has decided to introduce specific scenarios about the irrigation of oil palm cultivation in addition to them regarding the other field operations: this decision has been supported by the literature which highlights how often the oil palm is cultivated in areas characterised by low precipitation levels where irrigation is needed to reach the expected yields of fruits despite the high economical and environmental costs. Therefore, the MCA has to focus on this phenomenon in a separate way during the planning phase when the most suitable areas must be chosen in order to optimise the sustainability of the palm oil production.

Moreover, the logistics has been hypothesised in all the applications developed in a simplified way: in two applications (see Chaps. 5, 6) only the transport distances are considered whilst the vehicles typology is taken into account for the woody biomass production (see Chap. 4) and the storages organisation is analysed for the energetic use of agricultural residuals. These different approaches are due to the specific peculiarities and critical aspects of each case study: particularly, it is necessary to highlight that the logistics is often identified as the most problematic phase of the agro-energetic chains mainly from the economical point of view because of the low territorial density of the biomass and the seasonal nature of its supply which creates the need for a temporarily stockpiling before and after the delivery to the power, heat or processing plant.

During the decision criteria identification phase, for each case study a limited number of criteria have been fixed considering only two sustainability categories: environmental and economical. In fact, the MCA may be carried out considering a large number of additional possible principles considering different aspects, e.g. social, logistic, healthy, etc. aspects, and usually the choice of the principles taken into account must be supported by the necessity to reach objectives at a local, regional or global scale. As previously highlighted these objectives may be promoted by specific groups of stakeholders, i.e. politicians, technicians, customers, farmers, manufacturers, etc. Therefore, for the proposed applications only the economical and environmental criteria have been considered because these two typologies of criteria may be considered the most important ones for a large number of stakeholders: for instance, the economical criteria are certainly important for both producers/farmers and customers because the right solution can allow to minimise the market costs assuring a good level of competitiveness for the first ones and a profitable opportunity for the second ones; similarly, the environmental criteria may be considered fundamental for politicians, producers and customers in order to guarantee the environment preservation and a higher level of the life quality.

Moreover, for each criterion a measurable indicator has been identified and for each indicator three different levels of suitability have been defined. In this way, three values, alphabetical or numerical, may be assigned to each indicator: particularly, in all the proposed applications at the first stage of the methodology implementation the high level, i.e. the most profitable, has been identified by the letter "A", the medium level by the letter "B" and the low level by the letter "C", whilst in a second phase these alphabetical values have been converted into numerical.

Only for two cases analyzed in this work in Chaps. 3 and 6, criteria weights have been implemented adopting the "intra criteria" method described in Sect. 2.2. Assuming that the environmental and the economical criteria are characterised by the same level of importance, different weights have been associated to each indicator comparing it with the others of the same set of criteria. Moreover, the weights are numerical and allow to calculate for each scenario the environmental and the economical sustainability separately as weighted means associating with each criterion the corresponding indicator value multiplied for its weight.

Fig. 2.7 Components of a product life cycle assessment according to ISO 14040

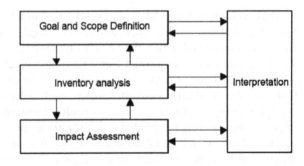

Fig. 2.8 Decision ranking matrix adopted for criteria scores combination

	A	B	C
A	A	A	B
B	A	B	C
C	B	C	C

The preference modelling has been carried out in two different ways: adopting alphabetical and/or numerical values during the scoring phase. As shown in Fig. 2.6 this phase has been first implemented in the proposed applications using the alphabetical scores and combining them through the decision ranking matrix illustrated in Fig. 2.8. Then the analysis of the obtained outputs has highlighted some criticism due to the large number of assumed criteria and reduced suitability level fixed for each criterion. Actually, the difference between the results of the scenarios, and mainly of the whole chains, has seemed insufficient determining a very limited number of chains at the "A" level despite a very large number of chains at the "B" level. For this reason the numerical scoring has also been applied by averaging the scores associated with each criterion on the number of environmental and economic criteria separately, and then making a final average between the two. The pros and cons due to the qualitative and quantitative approaches of the scoring phase have been compared more deeply in Chap. 5.

Once the MCA has selected the most suitable chains, the LCA has been applied to these alternatives in order to assess their environmental sustainability for a different point of view. As illustrated also in Chap. 3, for environmental evaluation, it is needed to distinguish between local and global pressures so that the agro-energetic or production chains may originate. For the first, it is necessary to take into account the principles on which the EIA is based; for the second, the LCA methodology is needed. Actually, during the last decades the relevance of the LCA in the food and agro-energetic sectors has been, respectively confirmed by the diffusion of different typologies of product certifications and by the recent indications of the EC (RED; [6]).

For each case study all the four phases of the LCA have been carried out according to the ISO standard and implemented all the inventory data in the software GEMIS 4.5 [7].

Particularly, the evaluation of the chains has been carried out comparing only two impacts calculated by the LCA: the GHG emissions through the CO_2eq and the primary energy consumption through the cumulative energy requirement (CER). The choice of these two indicators allows to evaluate the relevant impacts for all the proposed applications, even if they supply partial information about the environmental sustainability excluding other emissions on soil, water and air. Moreover, the CO_2eq emissions may be the first step for the environmental certification of the food products through the application of the Carbon footprint methodology, while they constitute the basis of the sustainability criteria proposed by the European Commission for solid and liquid biofuels (RED; [6]).

In this book, CO_2eq emissions have been calculated through the following formula using the GWP indices for only three chemical compounds, i.e. CO_2 with GWP = 1, CH_4 with GWP = 296, N_2O = 23, as proposed by the RED: Total *Global Warming Potential* [kg CO_2equivalent] = $\sum_i GWP_i.m_i$

The CER indicator is used to evaluate the overall energy consumption of fossil fuels during the production of a particular product. During the eighties this indicator has been used in the agro-industrial sector as a result of the energetic analysis [9], in order to measure the embodied energy in goods and services. In addition, this indicator is a quantitative indicator and allows to estimate not only the energy used directly in manufacturing or in supplying goods and services (direct consumption), but also the energy required to make available raw materials and equipment required for the production process.

The LCA methodology has been applied to each case study using the substitution method through the boundaries extension approach, in order to evaluate the role of the by-products: for this reason the LCA results shown in Chaps. 5 and 6 include some credits for the alternatives which suggest the reuse of the residues produced along the chain. No allocation has been done even if the proposed chain has concerned only the by-products reuse: this is the case of the agro-energetic chain illustrated in Chap. 3, where no impacts due to the olive-grove management for the olives production have been associated with the olive trees pruning, but only the emissions caused by harvesting, treatment, transport and energy utilisation.

Concerning the field phase the use of agricultural machines has been considered in terms of both diesel fuel consumptions (indirect emissions) and the associated direct emissions during the operations. Particularly, for diesel fuel it has been hypothesised an LHV of 11.86 kWh/kg and a density of 0.8 kg/l, while the quantity of diesel fuel required during the field operations has been estimated considering an average utilisation of 60% of the nominal power of the machine and a specific fuel consumption of about 0.20–0.25 kg/kWh [3, 5].

Finally, the transports of the products have been modelled in the LCA taking into account that usually local transports within the farm are carried out with tractors equipped with rural trailers, whilst greater distances provide the use of trucks.

References

1. Chiaramonti D, Recchia L (2010) Is life cycle assessment (LCA) a suitable method for quantitative CO_2 saving estimations? The impact of field input on the LCA results for a pure vegetable oil chain. Biomass Bioenergy 34:787–797
2. CIFOR (1999) AA.VV, Guidelines for applying multi-criteria analysis to the assessment of criteria and indicators, 9—The criteria & indicators toolbox series, published by Center for International Forestry Research (CIFOR)
3. Cini E, Recchia L (2008) Energia Da Biomassa: Un'opportunita'per Le Aziende Agricole. Pacini Editore, Pisa
4. Curran MA (2008) Life-cycle assessment, Human Ecol. 2168–2174
5. ENAMA (2005) Prontuario dei consumi di carburante per l'impiego agevolato in agricoltura. Roma
6. EC (2010) Report from the Commission to the Council and the European Parliament on sustainability requirements for the use of solid and gaseous biomass sources in electricity, heating and cooling, SEC 65-66 (2010). Brussels
7. GEMIS (2010) http://www.oeko.de/service/gemis
8. openLCA project. http://www.openlca.org/
9. Riva G (1996) I bilanci energetici: indicatori di processi industriali eco-compatibili, La Termotecnica, gennaio/febbraio, 33–45
10. Schlamadinger B, Apps M, Bohlin F, Gustavsson L, Jungmeier G, Marland G, Pingoud K, Savolainen I (1997) Towards a standard methodology for greenhouse gas balances of bioenergy systems in comparison with fossil energy systems. Biomass and Bioenergy 13: 359–375
11. Van den Berg NW, Duthilh CE, Huppes G. (1995) Beginning LCA: A guide into environmental life cycle assessment. Report 9453, NOVEM National Reuse of Waste Research Programme, Center of Environmental Research, Leiden
12. Wrisberg N, Udo de Haes HA,Triebswetter U, Eder P, Clift R (2000) Analytical tools for environmental design and management in a system perspective, CHAINET—European Network on Chain Analysis for Environmental Decision Support, Leiden, the Netherlands

Chapter 3
Energetic Use of Biomass and Biofuels

3.1 Introduction

Mediterranean countries produce 95% of the total world olive oil production
estimated to be 2.4 million tonnes per year. Olive production is a significant land
use in the southern Member States of the EU with important environmental,
social and economical implications. The main areas of olive oil production
are located in Spain (2.4 million ha), followed by Italy (1.4 million ha), Greece
(1 million ha) and Portugal (0.5 million ha). France is a much smaller producer,
with 40,000 ha.

In fact, data produced in 2000 by the EC's "Oliarea" survey [5] indicate a total
olive area in Italy of 1.4 million ha, which represents a 400,000 ha increase
compared with the area existing at the beginning of the 1990s and this is the
tendency observed during the last decade. Moreover, based on data of the 5th
Agricultural Census developed by ISTAT in 2000 and updated in 2005, the
Tuscany region is characterised by a land use which promotes olive-groves
occupying 18% of the total arable land.

However, in Italy there are several typologies of olive-groves because of different geographical area and site characteristics. Olive-groves may stronger differ
by trees density which may vary from 50 to 400 plants/ha, from older plantations
to the newer ones. Therefore, these different typologies of olive-groves require
very different management practices: the lower tree density, the lower frequency
of pruning operations of the trees crown.

The trees density also determines specific water requirements in order to assure
elevated yields and to limit the competitions between the plants. For this reason the
water availability is often considered as the most important limiting factor which,
in fact, determines the olive-groves characteristics. Literature [5] declares that
"under rain-fed cultivation, the lower the rainfall, the lower the tree density, by
necessity". The water supply through an efficient irrigation system may assure
sufficient yields even if the rainfalls are reduced, but the increases of the

L. Recchia et al., *Multicriteria Analysis and LCA Techniques,*
Green Energy and Technology, DOI: 10.1007/978-0-85729-704-4_3,
© Springer-Verlag London Limited 2011

Fig. 3.1 Manual pruning in a typical olive-grove (Azienda Belvedere, Castiglion della Pescaia, Grosseto)

production costs and of the environmental pressures may be carefully taken into account.

Pruning is an important aspect of olive-grove management. In fact, trees may be pruned every year or every two or more years, depending on local tradition, individual farmer decisions, etc. In Tuscany, in some marginal situations or in particular areas where some constraints for preserving the landscape are fixed, trees are not pruned for many years and develop high dense canopies: this practice could cause problems in terms of workers' safety during olive harvesting and could also significantly increase the costs (see lower yield and higher harvesting time per plant).

Pruning operations are usually effectuated manually with chainsaws (see Fig. 3.1) and require a large labour input (i.e. one-fifth to one-third of the total labour on traditional farms). Recently, new varieties that can be mechanically pruned and consequentially are able to reduce significantly the labour needed, have started to be cultivated. This is the case of the so-called "bush" varieties planted in dense rows.

The Literature highlights that pruning must be carefully taken into account in order to assure the health of the trees: it is known that regular pruning and the adoption of specific codes of practice are able to decrease drastically pest or fungi risks. In addition, a crown with limited size allows to reduce the amount of pesticides needed with minor environmental risks.

Nevertheless, pruning residues management is a complicated and expensive agricultural operation. In Italy, usually, farms manage olive tree prunings in three different ways: harvesting and burning in the headlands; shredding and landfilling; harvesting, conditioning and storing for energetic use.

Each of these techniques is characterised by both advantages and disadvantages [8], as briefly shown in Table 3.1.

In addition, it is possible to affirm that choosing how to manage olive tree pruning depends on several factors and the most important one is the economical sustainability of the whole agro-energetic chain: it must be assured that the biofuel production cost is convenient and comparable with those of fossil fuels. Considering that one of the most critical aspects of agro-energetic chains is the low

Table 3.1 Pros and cons for different management systems of pruning residues [22]

Management of olive tree pruning	Advantages	Disadvantages
Harvesting and burning in headland This operation is done in olive-groves where pruning is manually carried out	Branches with larger diameter (>4–5 cm) can be selected and used as biofuels in wood stoves or fireplaces	High quantity of labour is required in order to move residues from field to the headland Several municipal laws forbid to burn pruning and other residues in field because of fire risk and air pollution
Shredding in field and landfilling The pruning residues are windrowed, shredded and then landfilled or left in field	No transports are necessary No areas must be dedicated for temporal storage of pruning residues Pruning residues are able to increase the organic matter in the soil because in a few months they are completely mineralized	Residues can be a diffusion means of several pests from tree to tree High quantity of labour is required
Harvesting and energetic use The pruning residues are windrowed, shredded and used as biofuel in an energy production plant	Environmental benefits in terms of CO_2eq savings Economical benefits in terms of reduction of the use of fossil fuels Partial independence from traditional fuels	Logistics is complex Storage areas must be provided

density of the biomass on the territory, which implies low operative capacities during the harvesting phase and significant transport distances, it is important to correctly evaluate the available quantity of pruning for specific areas of interest.

Taking into account all the existing varieties of management and also all the possible differences in site characteristics (with particular attention to land extension, plant yield, agronomical characteristics of the variety, soil fertility, etc.), pruning density may vary a lot.

The Literature reports several methodologies to estimate the biomass quantity, as the empirical equation developed by the ANPA [2] and reported in the following

olive tree pruning on dry basis (t_{db}/ha/year) = 0.566 × yield (t/ha/year) + 1.496.

According to the formula, Ambiente Italia [1] indicates quantities of about 1.0–4.0 t_{db}/ha/year, whilst ITABIA [17] suggests values of 0.2 and 0.8 t_{db}/ha/year for wood and foliage of olives, respectively.

Moreover, tests carried out by the University of Florence during the previous years allowed to collect data for various areas of the Provinces of Firenze, Grosseto and Siena:

- Experimental harvesting conducted at the Montepaldi farm in the Province of Firenze, has detected an average density on wet basis of 5.0 t_{wb}/ha/year [18];

Table 3.2 Physical and chemical characteristics of olive tree pruning [9]

Parameters	Units	Standard values	Standard ranges
Moisture	%	47.0 (wood); 62.0 (foliage)	45.0–50.0 (wood); 60.0–65.0 (foliage)
LHV	MJ/kg_{db}	19.0	18.4–19.1
C content	$%_{db}$	52.0	50.0–53.0
H content	$%_{db}$	6.1	5.9–6.3
O content	$%_{db}$	41.0	40.0–44.0
N content	$%_{db}$	0.5	0.3–0.8
S content	$%_{db}$	0.04	0.01–0.08
Cl content	$%_{db}$	0.01	<0.01–0.02
Ca content	$%_{db}$	4,000	3,000–5,000
Mg content	$%_{db}$	250	100–400
Na content	$%_{db}$	100	20–200
P content	$%_{db}$	300	–
Si content	$%_{db}$	150	7–250
Cd content	$%_{db}$	0.1	–
Hg content	$%_{db}$	0.02	–
Pb content	$%_{db}$	5.0	–
K content	$%_{db}$	1,500	1,000–4,000
Ash content	$%_{db}$	1.7 (wood); 6.0 (foliage)	1.5–2.0 (wood); 5.0–7.0 (foliage)

- Close to Grosseto, multiple sampling has been carried out [7] measuring a density of 2.0–3.0 t_{wb}/ha/year;
- Tests executed at the farm Castello di Fonterutoli in the Province of Siena have highlighted values much lower than the previous one ranging between 0.8 and 1.6 t_{wb}/ha/year for 1- and 2-year pruning, respectively [8].

Therefore, in this work an indicative density on wet basis of 1 t_{wb}/ha/year for 1-year pruning has been fixed.

Finally, once the available quantity has been estimated and the use of pruning as biofuel has been chosen, it is useful to make some considerations about the energetic characteristics of these residues. Table 3.2 reports the physical and chemical characteristics of olive tree pruning. These values may vary if some harvesting conditions occur as briefly indicated in Table 3.3, decreasing the performances (efficiency, durability, etc.) of the energy plant.

In any case, for the present work a lower heating value (LHV) of 11.3 MJ/kg_{wb}, equal to 3.154 kWh/kg_{wb}, has been established.

3.2 Goals, Definitions and Decision Makers' Typology

The aims of the present work are:

- Defining all the scenarios identifying the possible harvesting yards, transport conditions and energetic plants;

Table 3.3 Possible causes of variation of some pruning characteristics [9]

Characteristic variation	Possible causes
Higher ash content	Earth contamination
	Higher content of bark
	Inorganic added compounds
	Chemical treatments
Lower/higher LHV	Presence of material with lower/higher LHV
Higher content of N	Higher content of bark
	Content of plastic materials mixed with
Higher content of S	Higher content of bark
	Inorganic added compounds
	Chemical treatments
Higher content of Cl	Higher content of bark
	Wood coming from coastal zones
	Contamination with salt (i.e. salt used on roads in winter time)
	Chemical treatments (i.e. compounds used to preserve the wood)
Higher content of Si	Earth contamination
	Higher content of bark
Higher content of Ni	Contamination through machine used for conditioning
	Mineral oil
Higher content of Pb	Environmental contamination (e.g. vehicle traffic)
	Content of plastic materials mixed with
	Traces of painting

- Assessing the effective implementation of the agro-energetic chains that can be obtained as a combination of all the scenarios, excluding those evaluated as impossible;
- Defining the sustainability from the environmental and economical points of view through some specific criteria previously fixed that allow to indicate the more suitable agro-energetic chains;
- Carrying out a more detailed environmental analysis able to investigate additional aspects according to standardized indicators, in order to choose the chain with lower environmental impacts.

As discussed in Chap. 1, all this methodology can be applied at the planning stage or during the development of the feasibility study. Therefore, it is possible to identify as decision makers politicians, technicians and/or farmers basing on the different stages of the decision step.

3.3 Scenarios Definition

Agro-energetic chains based on the reuse of agricultural residues are usually complex and characterised by various steps such as harvesting, processing into biofuel, packaging, transport.

Except from the harvesting phase, all other processes can be arranged in different ways. Particularly, the localisation and the timing of the operations can vary: for instance, the biomass conditioning may occur during the harvesting in field or may constitute a separated phase after the harvesting near the field, in a dedicated area or close to the energy production plant.

Therefore, in order to define all the possible scenarios, at first different solutions for each phase must be considered; then, combining together all the solutions previously identified, several scenarios must be set [19].

In the following, the main implemented solutions are illustrated regarding harvesting, conditioning, packaging, transport and energy conversion, analysing the field operations (i.e. yard characteristics), the logistics (i.e. biomass management and transport conditions) and the energy production phase (i.e. plant typology and its technical characteristics). It is important to highlight that each single choice in a specific phase may affect all the next ones: in fact, the decision concerning the typology of the biofuel produced (e.g. wood logs instead of wood chips) obviously influences the choice of the energy conversion plant discarding all the incompatible solutions.

3.3.1 Description of Yards

Regarding biomass harvesting, several possible yards have been defined taking into account usual olive-grove management, typology of operations currently carried out, level of mechanisation, etc. The choice of the most suitable yard for a specific case depends on various factors, such as ground conditions and slope, olive-grove organisation and management, headland size, etc. For instance, if it is needed to operate in terraces, only the more compact machines can be used; if slope is very high (>20%) no mechanisation is possible; whilst, on flat or moderately sloped terrain it is possible to adopt all sorts of equipment. Great attention must also be given to the presence of stones, which can be harvested together with the pruning residues and destroy some mechanical devices; therefore, if the terrain presents this characteristic, it is profitable to use a machine with pick-up utilities.

Moreover, it is also important to investigate the other environmental aspects influenced by the use of machines: for example, if in a specific area the hydrological risk is high, machines with reduced weight must be preferred with the aim to limit the risk and to avoid a more significant soil compaction.

Generally, olive-groves have distance between the trees rows which does not limit the choice of the size of the machines used. However, the presence of areas along the field may be useful to move and temporally store the harvested biomass, allowing a different organisation of the yard and a different choice of the equipment.

In some cases, it is convenient to know which type of management is adopted by the farmer, i.e. the cutting modus and frequency and/or pesticides use, because this information can advise about pruning in terms of diameter dimensions, density per square metre, foliage presence, chemical compounds contained in wood, etc.

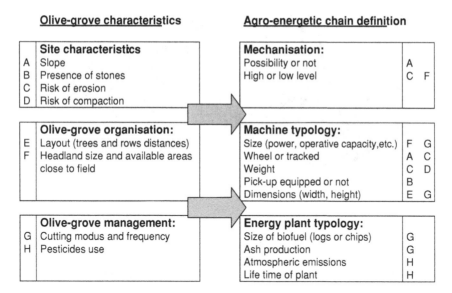

Fig. 3.2 Influence of olive-grove characteristics on different aspects of the agro-energetic chain. A letter (from A to H) has been associated with each characteristic with the aim to highlight by which of these characteristics each chain aspect is influenced

Particularly, the pruning size conditions the yard mechanisation and the harvesting technology: usually, small shredders are able to process branches with a maximum diameter of 5 cm, whilst industrial or forestry chippers are suitable for larger dimensions. In this way, the operative capacity of the yards and the biomass production costs are also fixed.

A consistent presence of foliage in relation to the wood mass (more than 25%), which is usually due to a low frequency of cutting (once a year), influences the reuse of the biomass: if the wood is used in a direct combustion plant, the combustion process will be affected by this characteristic and, for instance, corrosion phenomenon, atmospheric emissions of related compounds and ashes production will arise.

Accurate knowledge of the chemical product used for pest control is useful, even if no reliable experimental tests have been carried out for the absorption of these chemical compounds in wood but only for fruits in order to achieve a sufficient safety level for human health.

Finally, the arrangement of pruning plays an important role in the success of the harvest: to facilitate the work of machines, reduce losses and avoid clogging, pruning residues should be windrowed, taking into account the dimensions (width and height by the ground) of the harvesting machines.

In the scheme below the main olive-grove characteristics that contribute to define the harvesting yards are summarized: particularly, a letter (from A to H) has been associated with each characteristic with the aim to highlight the characteristics that influence the aspects of the agro-energetic chain (Fig. 3.2).

Fig. 3.3 Agricultural shredder equipped with a packaging system in olive-grove

On the basis of the literature [4, 8, 18, 21, 22] it is possible to identify four different techniques and yards to harvest, process and recover olive tree pruning: shredding in the field, shredding or chipping in the headland, baling and industrial harvesting.

Shredding in field is an interesting technique that allows to move the woody biomass from the field in a simplified way, reducing transport costs and improving logistics. Shredders result as a modification of traditional mulchers (see Fig. 3.3), equipped with a device able to collect the shredded residues with a volume variable between 2.0 and 7.0 m^3.

In general, machines have a box or a canvas bag (i.e. big bag) where the processed biomass can be temporally stored or they are able to deliver the product in a trailer alongside. These two different solutions must be carefully evaluated considering site characteristics, because the second requires larger operative areas and lower slopes.

In any case, the required power varies from 40 to 70 kW depending on the model; the maximum diameter treated is about 5 cm. Some machines are equipped with a pick-up: this additional utility allows to work on stony ground and to preserve the shredded wood from any contamination of grass and/or earth. It is possible to achieve an operative capacity of 0.6–0.9 t_{db}/h, with an average cost of 60 €/t_{db}.

Industrial shredders (see Fig. 3.4) are also available on the market; they are obtained by modifying disk or drum chippers, require high power (e.g. 150 kW) and allow to treat branches also with a diameter of 10–15 cm, reaching operative capacities of about 3.0–5.0 t_{db}/h and a corresponding cost of about 40 €/t_{dm}.

Shredding or chipping in the headland implies a traditional organisation of the operations before the harvesting phase: in fact, when pruning residues are not reused as biofuels, they are mulched and left in field or alternatively they are windrowed, moved with tractor equipped with fork (see Fig. 3.5) and finally concentrated in the headland where usually they are burned. Therefore, if shredding or chipping in the headland is proposed, this phase replaces the traditional burning of the residues and all the previous processes are carried out in a similar way.

Fig. 3.4 Industrial shredder in olive-grove. From *left* to *right*: delivery of the shredded pruning residues in the rural trailer, machine alimentation, processed biomass

Fig. 3.5 Tractor equipped with fork used to move the pruning from field to the headland

Considering the phase of pruning collection and storage in the headland, an operative capacity of 0.5–0.7 t_{db}/h and a cost of about 60 €/t_{db} may be hypothesised. For the next step the shredding or chipping machine powered with a hydraulic crane is operated by a tractor with a power of 100–120 kW. The operative capacity reached is 2.0–3.0 t_{db}/h with a cost of about 30 €/t_{dm}.

Baling is a processing technique suitable for the thin woody residues difficult to manipulate otherwise. This solution allows to organize the biomass into homogeneous units, and consequently facilitate handling and storage. Machines present on market can be divided into three groups: small rectangular balers, small round balers and industrial balers.

The small rectangular balers are derived by press-fodder machines, packing parallelepiped bales with a plunger device with reciprocating rectilinear motion. They are applied to an agricultural tractor with a power ranging between 40 and 60 kW, able to work on a front of 1.0–1.5 m. Bales have an average size of 45 × 35 × 70 cm; their weight is about 40 kg depending on the moisture of the olive tree pruning collected. The operative capacity is also variable because of several factors (i.e. site characteristics; type, size, moisture of residue; work conditions; etc.) but hypothesising a team of two workers (one driving the tractor and one facilitating the collection with a fork) it may be estimated about

600–1,000 bales/day correspondent to a 1.0 ton of wood on dry basis processed per hour with a production cost of 50 €/t_{db}.

The small round balers actually use the same operating principle of the industrial models, but have a limited size and weight (about 25% of the standard models) in order to be used in fields with a very reduced tree spacing. They can be applied to a small tractor able to supply a power of 25–30 kW. Bales can weight from 30 to 40 kg, depending on the type of material. These machines need only one operator and reach an operative capacity of about 1.6 t_{db}/h, compared to an estimated cost of about 25 €/t_{db}.

Industrial balers used to collect olive tree pruning are also derived from modified farm equipment. In this case the field of application is quite different: these machines can easily and efficiently work only in modern and well-organized plants, because their large size and the yard implemented require operative areas with adequate dimensions.

The diameter of the bales is equal to 1.0–1.5 m with a volume of 1.0–2.0 m^3, depending on the model; the weight per bale varies from 500 to 700 kg depending both on machine model and setting and on type of material collected. All the packaging functions are controlled by a computer directly used by the tractor driver which is able to do all the work alone. These machines can be applied to a tractor of 60 kW and reach an operative capacity of 2.0–4.0 t_{db}/h, with an average production cost of 20 €/t_{db}.

When all the biomass has been harvested and packaged, the bales must be collected from the field, usually through a tractor equipped with fork; in a second phase, the bales can be directly used in adequate kiln or processed in order to obtain a more performed biofuel (e.g. wood chips).

Finally, it is possible to implement industrial harvesting techniques, which allow to perform more processes in only one step and reduce drastically the associated costs.

Theoretically, the same machine can simultaneously perform the four operations required to recover the pruning residues, i.e. cutting, windrowing, conditioning and transport. Commercially, there are already some interesting solutions. For example, round balers equipped with some braches able to windrow, with a small trailer to store and move the bales and eventually also with pruners only in plantations where it is advisable. Another possibility is constituted by shredders (see Fig. 3.6) provided with a multidisc bar applied to a hydraulic arm on the right side of the machine, with a tank mounted on the front under the bar where a conveyor sends the pruning to the shredding chamber, with a grid of calibration to improve product quality and with a posterior tank with a capacity of 10 m^3 where shredded pruning is temporally stored.

Taking into account all the possible yards applicable to the pruning harvesting, it must be highlighted that logistics plays an important role: biomass collection, transport and delivery to the energetic plant must be efficiently organized. In fact, it is possible to implement different scenarios to optimize the biofuel production and management:

Fig. 3.6 Industrial shredder in olive-grove able to cut, collect, shred and transport the biomass simultaneously. On the right a particular of an olive tree branch after mechanical cutting

- Biomass chipping in field and wood chips delivery to the energetic plant;
- Pruning harvesting and concentration in the headland, biomass chipping in the headland or in other dedicated areas even after a natural drying period, chips use in the conversion plant;
- Bales production and transport to the user, eventual chipping of the bales (this phase can also be avoided if bales are directly used in the burner, i.e. cigar burner), use of the biomass in an energetic plant.

All these scenarios can be realized with several technologies available on the market, at industrial or semi-industrial level, according to different investment costs and different working performances (see Table 3.4).

Yards with high mechanisation levels are more suitable for specialized olive-groves characterised by a management which usually originates a larger quantity of pruning residuals than obtained in traditional plants. In this case the harvesting is carried out in a more efficient way using machines designed for the forestry sector than agricultural equipments. All these solutions present comparable costs for biomass harvesting and treatment only if solutions are correctly chosen considering site characteristics, olive-grove organisation and management. On the other hand, olive-grove with high extension (no field fragmentation) and regular layout with high distance (5–6 m) between rows permit the use of machines with a very high operative capacity able to reduce significantly the production costs.

On the basis of all the previous considerations, eight different scenarios have been fixed, as reported in Table 3.5. For each scenario a brief description of the harvesting yard indicating the machines used, the typology of the produced biomass and the associated costs have been listed. All the processing costs have been referred to a ton of wood on wet basis, considering an average moisture of 35% and have been collected in several references.

All the proposed scenarios provide a previous phase of manual pruning usually followed by windrowing (except for the scenario Y3): Fig. 3.7 illustrates these operations.

Table 3.4 Standard yards for the energetic reuse of olive tree pruning

	Maximum diameter of branches (cm)	Product	Power of tractor (kW)	Yard of wood harvesting	Yard of biofuel collection	Operative capacity (t_{db}/h)	Harvesting cost (€/t_{db})
Shredder in field							
Shredder from modified mulcher	5	Wood chips	40–70	Tractor with shredder One worker	Tractor with trailer One worker	0.6–0.9	55–60
Shredder from modified chipper	10–15	Wood chips	150	Tractor with shredder One worker	Tractor with trailer One worker	3.0–5.0	35–40
Shredder or chipper in the headland							
First phase: pruning collection	–	Wood branches	30–40	Tractor with fork One worker	–	0.5–0.7	50–60
Second phase: wood chipping	20	Wood chips	100–120	Tractor with chipper One worker	–	2.0–3.0	30–35
Baler							
Small rectangular baler	7	Parallelepiped bales, with a size 45 × 35 × 70 cm and a weight of 40 kg for each	40–60	Tractor with baler Two workers	Tractor with trailer Two workers	1.0	45–55
Small round baler	4–5	Round bales, with a weight of 30–40 kg for each	25–30	Tractor with baler One worker	Tractor with trailer Two workers	1.6	20–25

(continued)

Table 3.4 (continued)

	Maximum diameter of branches (cm)	Product	Power of tractor(kW)	Yard of wood harvesting	Yard of biofuel collection	Operative capacity (t_{db}/h)	Harvesting cost ($\text{€}/t_{db}$)
Industrial baler	4–5	Round bales, with a diameter of 1.0–1.5 m, a volume of 1.0–2.0 m^3 and a weight of 500–700 kg for each	60	Tractor with baler One worker	Tractor with fork Tractor with trailer Two workers	2.0–4.0	15–25

The operative capacity and the harvesting cost are referred to the woody biomass on dry basis

Table 3.5 Scenarios hypothesised for harvesting and processing the pruning residues in the MCA

Scenarios	Yard mechanisations	Biomass outputs	Biomass treatment costs (€/t$_{wb}$)	References
Y1	Windrower Tractor with fork	No biomass produced (biomass burned)	59.00	[4, 8]
Y2	Windrower Shredder Landfilling	No biomass produced (biomass landfilled)	50.00	[8]
Y3	Shredder with packaging system Tractor trailer	Wood chips	68.00	[8]
Y4	Windrower Shredder Tractor trailer	Wood chips	54.00	[4, 22]
Y5	Windrower Tractor with fork Chipper Tractor trailer	Wood chips	50.50	[4, 22]
Y6	Windrower Baler Tractor trailer	Square bales	13.50	[4, 22]
Y7	Windrower Baler Tractor trailer Chipper at storage	Square bales	45.50	[4, 22]
Y8	Windrower Baler Chipper at storage	Round bales	65.50	[4, 22]

Fig. 3.7 Pruning and windrowing in a Tuscany olive-grove close to Grosseto: manual activities and machine adopted

Particularly, scenarios Y1 and Y2 hypothesise that there is no energetic reuse of the pruning residues. In the first one the biomass is collected in the headland and then burned, even if a large number of Municipalities in Italy forbid this activity through specific laws and regulations. The second one provides the shredding of the pruning residues afterwards left in field and landfilled: this practice is profitable in order to preserve and in some cases increase the organic matter of the soil and the amount of nutrients, but often it is not implemented because of the risk of pests diffusion from tree to tree.

In any case, even if these two scenarios imply a low level of mechanisation, they are not free of charge: the associated costs are similar to those estimated for the other solutions and may also increase for specific site characteristics (i.e. field fragmentation, high slope, low pruning density, reduced operative areas along the field, etc.). Regarding the scenario Y1, it must be highlighted that the biomass treatment cost largely depends on the movement of the wood branches from the field to the headland (see Table 3.4).

Scenarios Y3 and Y4 concern yards able to shred the biomass in field and they only differ for few aspects. Y3 hypothesises the use of a shredder equipped with a packaging system, i.e. big-bags, used for wood chips storage and delivered on field when fulfilled, whilst Y4 adopts a machine which ejects the processed biomass in a rural trailer following the shredder during its activity.

Moreover, Y3 does not require a windrowing phase before harvesting starts, because it is supposed that the machine used here can intercept the pruning passing more than one time into the same row adopting a speed higher than usual. This way to operate is also profitable in order to limit the height of the pruning piles and consequently avoid stops due to machine clogs.

The Y5 scenario also produces wood chips but in this case the biomass is firstly collected using a tractor with fork and then chipped in the headland. In this case, the cost results as lower because the chipper has an operative capacity higher than those of the small shredders adopted in scenarios Y3 and Y4, and because this organisation of the yard allows to avoid or significantly reduce all the additional times (i.e. extra times which must be added to the working time of the machine) due to machine reversing, stopping for technical controls and/or resetting, etc.

Finally, the last three scenarios provide the use of different types of baling machines: in this case, even if the processing costs and the mechanisation level seem lower than those hypothesised for the other scenarios, it must be considered that biomass will be almost certainly chipped in a subsequent phase in order to be used in an energy plant. The Y6 scenario where small rectangular bales are produced presents a very low cost, but whether bales must be chipped in an intermediate phase or can be directly used in a combustion plant with low power, additional costs, higher than these for the other two scenarios, must be taken into account: in the first case chipping will require more extra time due to the alimentation phase and this extra time will be much higher as the bale sizes will be reduced; in the second case the boiler will have a lower efficiency during combustion and much more wood will be required to supply a certain quantity of thermal energy.

Table 3.6 Scenarios hypothesised for logistics in the MCA

Scenarios	First transport (km)	First storage	Second transport (km)	Second storage	Plant typology
L1	0	–	0	–	–
L2	15	Farm storage	0	–	Small-sized
L3	15	Farm storage	40	Plant storage	Medium-sized
L4	15	Farm storage	100	Plant storage	High-sized
L5	0	–	40	Plant storage	Medium-sized
L6	0	–	100	Plant storage	High-sized

In conclusion, all the proposed solutions can be easily implemented in the farms deciding, from time to time, the mechanisation level (and consequently the typologies of machines to be used) and the biomass output. Moreover, the processing costs seem to be comparable with one another and also with solutions that do not provide the reuse of olive tree pruning: this fact is very important to promote the biofuel production from residual materials instead of dedicated crops without causing additional environmental costs (external costs).

3.3.2 Description of Logistics

One of the most important problems in using biomass as a fuel is the spreading out of supplies together with the low territorial density, in comparison with the traditional fossil fuels. In most of the cases the biomass supply is seasonal, namely variable in time, thus creating the need for a temporary stockpiling before and after delivery to the power, heat or processing plant. In addition, storage costs are mainly due to the management of great volumes of biomass, since its low specific weight and need to be stored in large quantities. On the other hand, transport costs are in general mainly dependent on geographic issues [6].

On the basis of these assumptions, different scenarios for biomass transport and storage have been designed. The hypothesis developed is mainly generic because of uncertainty about typology, size and localisation of the energy conversion plants. Particularly, the following scenarios have been defined (see Table 3.6):

1. Scenario L1 does not identify any storage as when pruning are burned along field or landfilled;
2. Scenario L2 hypothesised a reuse of the biomass within the farm for thermal energy production in a small-sized plant. It is important to highlight that this solution does not assure the effective wood utilisation in an appropriate conversion plant characterised by high efficiency, low atmospheric emissions, etc. In fact, farmers could use these biomass in stoves, fireplaces or in heat boilers with an efficiency lower than 70% and without any control technology for the atmospheric emissions;

3. Scenarios L3 and L5 are referred to a medium-sized energy plant which is supposedly located within a distance of 40 km. Particularly, in the first scenario it is also provided an intermediate storage, whilst in the second the biomass is directly delivered to the energy plant;
4. Scenarios L4 and L6 concern high-sized energy plants sited at a distance of about 100 km, respectively, with an intermediate storage at the farm or not.

It is important to highlight that these possible scenarios assume only two different typologies of means of transport: for the biomass movement within the farm a tractor equipped with a rural trailer is used; for the biomass delivery to the plant it is also necessary to utilise highway trucks or similar vehicles. Obviously these two typologies of means have very different characteristics in terms of load capacities, speed, atmospheric emissions; therefore, they imply different environmental and economical performances which may be evaluated. Usually, rural transports imply higher costs due to lower speed and load capacity of the mean; whilst trucks cause higher atmospheric emissions. Moreover, limiting the transport distance assures the diffusion of pollution (i.e. atmospheric emissions and noise) only within a constrained area.

3.3.3 Description of Energy Plants

Concerning the energy utilisation scenarios, different plant typologies have been considered: small-, medium- and high-sized direct combustion plants, able to use the woody biomass with different characteristics (i.e. wood chips or bales).

In general, plants using small bales as biofuels are characterised by low power and no automation for the alimentation phase, whilst wood chips boilers may be small-, medium- or high-sized. A significant exception is represented by high-sized plants which are able to burn large round bales without providing a previous phase of chipping: they are mainly used for herbaceous biomass and equipped with specific burners called "cigar-burners"; they are diffused in Northern Europe, but in Italy they are not yet present.

In Table 3.7, the main typologies of energetic plants for biomass are listed, indicating also some characteristics of the biofuels used in [14].

Moreover, besides price, the main characteristics of a boiler are its power (kW), efficiency (%), service life (years) and user-friendliness or level of automation. In the last 25 years, wood-fed boilers have undergone significant technological improvement.

The highest technological level is observed in small- and medium-sized automatic chip-fed boilers, where there is no need for the presence of a person to manually stoke the fuel.

In the 1980s the average efficiency of wood-fed boilers was about 50–60%, whereas today it exceeds 80–85%, reaching over 90% in the best models. As a positive consequence, emissions have significantly decreased and reliability and comfort have been raised: Table 3.8 shows standard levels of VOC and NO_x emissions for domestic plants which usually are not equipped with specific

Table 3.7 Plants typology for the energetic reuse of the olive tree pruning

Plant typology	Range of power (kW)	Plant automation	Biofuel characteristics		
			Typology	Ash content (%)	Moisture (%)
Wood stove	2.0–10.0	No automation	Wood logs	<2	5–20
Wood boiler	5.0–50.0	No automation	Wood logs	<2	5–30
Pellet stove and boiler	2.0–25.0	Partial automation	Pellets	<2	8–10
Boiler with under stocker furnace	20–2,500	Total automation	Wood chips and residuals	<2	5–50
Boiler with travelling grate furnace	150–15,000	Total automation	Biomass	<50	5–60
Boiler with grate furnace and preheating system of biofuel	20–1,500	Total automation	Woody residuals	<5	5–35
Boiler with under stocker furnace and rotating grate	2,000–5,000	Total automation	Wood chips	<50	40–65
Boiler with cigar burner	3,000–5,000	Total automation	Round bales	<5	20
Boiler with whole bales furnace	12–50	Total automation	Small bales	<5	20
Boiler with fixed bed furnace	5,000–15,000	Total automation	Biomass	<50	5–60
Boiler with circulating fluidised bed furnace	15,000–100,000	Total automation	Biomass	<50	5–60
Boiler with dust firing furnace	5,000–10,000	Total automation	Biomass	<5	20

Table 3.8 Atmospheric emissions of domestic plants for woody biomass [20]

Plant typology	VOC (mg/kWh)	NO_x (mg/kWh)
Traditional wood boiler	1,000	350
Modern wood boiler	300	520
Modern wood stove	700	n.a.
Wood pellet boiler	160	<270
Wood pellet stove	120	<270

abatement systems because of the reduced size. These data are particularly important because airborne emissions constitute a critical aspect mainly for small-sized boilers where low temperatures of combustion, insufficient and non homogeneous air–fuel mix, and frequent stops and starts of the plant can cause emissions about ten times higher than those of industrial plants.

All different scenarios considered in the MCA are described in Table 3.9, reporting also additional information [3, 13, 16, 17] as annual fuel demand, annual

Table 3.9 Scenarios hypothesised for energetic plants in the MCA

Scenario	Plant typology	Operating time (h/year)	Efficiency (%)	Investment costs (k€/kW)	Plant automation
E1	Small-sized (i.e. 20–60 kW) direct combustion plant using small bales for heat production	2,200	70	0.35–0.40	No automation
E2	Small-sized (i.e. 100 kW) direct combustion plant using wood chips for heat production	2,200	85	0.40–0.60	Partial automation
E3	Medium-sized (<1 MW) direct combustion plant using wood chips for heat production	2,200	85	0.15–0.20	Partial automation
E4	Medium-sized (i.e. 1 MW e 150 kW e) direct combustion plant using wood chips for electricity production	5,000	15	3.50–5.00	Complete automation
E5	High-sized (i.e. 10 MW e 3.6 MW e) direct combustion plant using wood chips for electricity production	7,000	20	5.00–7.00	Complete automation

operating time, energy efficiency and investment costs. Particularly, the scenario E1 identifies an energy plant which uses directly the small bales of the harvested pruning, while the other scenarios hypothesise the use of wood chips. In addition, it must be highlighted that the thermal energy production has been preferred and only two scenarios (E4 and E5) adopt plants able to produce electricity through gas turbines; this is also shown by higher annual operating times, lower efficiency values, higher investment costs and higher level of automation of the plant. In fact, woody biomass is more suitable for thermal energy production, also taking into account that power plants present higher size and consequently require higher amounts of biofuels, complicating the relative logistics, implying elevated transport distances and finally determining high environmental pressures and management costs for biomass supply.

3.4 Decision Criteria Identification

For each scenario of biomass harvesting (Y scenarios), logistics (L scenarios) and energy utilisation (E scenarios), some decision criteria and indicators concerning environmental impacts and economical sustainability, have been defined as reported in Table 3.10.

Table 3.10 Definition of evaluation criteria and indicators for bio-energy scenarios (i.e. Y = biomass harvesting and treatment; L = logistics; E = energy utilisation)

Criteria and indicators	Scenario typology	Assigned values		
		$A = 1$	$B = 2$	$C = 3$
Environmental aspects:				
No. of machines used during field operations	Y	1–2	3	4
Distance of transport (km)	L	≤50	>50 and ≤100	>100
Efficiency of the energetic plant (%)	E	≥85	≥50 and <85	<50
Economical aspects:				
Biofuel production cost (€/t)	Y	≤50	>50 and ≤60	>60
No. of storages	L	0	1	2
Plant investment cost (k€/kW)	E	≤0.3	>0.3 and ≤2.0	>2.0
Plant management	E	Complete automation	Partial automation	No automation

Particularly, the decision criteria have been determined taking into account several aspects.

From the environmental point of view, scenarios can be preferred if the number of machines involved for harvesting is reduced and mechanisation level is low, if transport distances are limited, if the energy plant has high conversion efficiency. Moreover, economical benefits are possible if the costs of biofuel production and treatment are limited, if the storage operations are reduced, if the investment costs for the energy plant are limited and if the management of the energy plant is easy with high level of automation (low number of workers, few specialized workers needed, etc.).

Concerning the proposed environmental criteria it is necessary to explain the choices made. The methodology described in this book provides as a first step the application of an MCA and, only in a second step, of a more specific approach with the aim to detect additional environmental pressures. As illustrated in Chap. 1 the second step is carried out implementing the LCA and measuring only two environmental impacts, the CO_2 equivalent and the CER for the more suitable scenarios identified through the MCA. These two indicators are very important for the agro-energetic chains hypothesised which provide the reuse of the olive tree pruning as biofuels, because they allow to evaluate the performances of the different scenarios at a global level in terms of GHG savings considering also the energy efficiency. In contrast, this approach must be considered as a complementary tool of the assessment carried out through the MCA where other impacts on the territory have been considered in order to evaluate the effects of the described operations at a local scale.

In fact, usually the environmental impacts of agriculture at a local level can be evaluated adopting an Environmental Impact Assessment (EIA) and selecting a specific group of pressure indicators which are ideally assessed for each environmental zone. The main pressure indicators used are erosion, soil compaction, nutrient leaching to groundwater and surface water, pesticide pollution of soils and water, water abstraction, fire risk, biodiversity [11]. In addition, for olive-groves it is also important to consider the possible effects related to landscape modifications.

Different pruning managements can indirectly modify these pressures. For instance, concerning pruning management it is possible to highlight that

- Reduced mechanisation of the harvesting phase may limit the compaction and erosion problems;
- Chipping and landfilling the pruning residuals is a good technique to increment the organic matter in soil, increasing its fertility and reducing the erosion;
- Limiting the soil compaction allows a better infiltration of the water in the soil, reducing the run-off phenomena and irrigation needs (trees are able to use the water absorbed by the ground);
- Burning of pruning residues causes some air pollution and fire risk.

Besides, all these assumptions have been indirectly considered when the environmental criteria of the MCA have been fixed. In fact, it is possible to affirm that a reduced number of machines corresponds to a low mechanisation level and consequently assure a lower risk of erosion and compaction of the soil, which is affected by lower machine weights and passages. In this way it is also possible to increase the possibility of grass and weeds diffusion which are able to contrast these phenomena and promote a biodiversity preservation. Moreover, avoiding the compaction of the soil guarantees a better absorption of rainfall and consequently reduces irrigation needs, i.e. water consumption. In addition, a limited transport distance implies shorter trips of transport means and so lower pollution due to atmospheric emissions and noise affect the territory.

Finally, for the energy plants the global efficiency has been taken into account: this choice is based on the indications of the recent guidelines of the EC about the environmental sustainability of solid biofuels [10]: it is important to know how the biofuels are used during the conversion energy phase because the higher the energy efficiency of the conversion, the higher the fossil fuel savings and the lower the required quantity of the biofuel and the impacts due to its production.

Taking into account a planning stage or developing a feasibility study, the costs analysis can be done with a very limited accuracy level. In general few data are available in these phases, therefore several information collected by the literature are used avoiding to develop detailed calculations. Table 3.10 reports as a first criterion the evaluation of the biofuel production cost, assuming that the best values are those less than 50 $€/t_{wb}$. This limit is imposed by the market where wood chips are sold at about 30–40 $€/t_{wb}$ at the high-sized energy plants.

Moreover, also the storage cost must be considered because not only logistics is more complicated but also indirect costs are required: for example, additional biomass treatment may be needed in order to guarantee the biomass conservation for a longer period (i.e. some woody biomass with high mois-ture content can be previously dried reducing the risk of fermentation phenomenon).

Regarding the conversion phase, it is necessary to specify that only investment costs of the energetic plants have been considered, because usually in Italy one of the main barriers against diffusion of the biofuels is the initial investment even if it is possible to use public financing and/or taxes reduction.

In any case also the management costs of the plants have indirectly been considered: in fact the higher the level of automation, the lower the costs.

3.5 Values and Weight Associated with the Criteria

For each criterion a specific weight has been fixed, as reported in Table 3.11.

The weights are numerical and allow to calculate for each scenario the environmental and the economical sustainability separately as weighted means associating to each criterion the corresponding value (see Sect. 3.4) multiplied for its weight.

No suggestions are supplied in order to determine which sustainability is more important between the environmental and the economical one, because they are not comparable from a technical point of view but only from a political one. If decision makers or planners decide to promote one sustainability in respect to the other an additional weight can be introduced to differentiate the contribution of these two factors within the calculation of the global value.

Between the environmental criteria and the associated indicators the efficiency of the energetic plants has been classified as the most important (two times more important than others) characteristic following the indication in [10]. Particularly, the EC decided to introduce this criteria according to the objectives of the RED where the utilisation of the renewable sources (+20% within 2020) has been contemporarily promoted with the increase in energy efficiency (+20% within 2020). On the other hand between the economical criteria and the associated indicators the biofuel production cost has been evaluated as the most significant one (two times more significant than others), because the literature [4, 6, 8, 13, 22] indicates that this aspect is the most critical in order to decide if the agro-energetic chain is convenient or not in respect to the energy production through fossil fuels.

3.6 Results of the MCA for All Possible Agro-Energetic Chains

For each scenario illustrated in Sect. 3.3 for harvesting, logistics and energetic conversion, the judgements have been calculated applying the environmental and the economical criteria, as reported in Tables 3.12, 3.13 and 3.14.

Particularly, it can be highlighted that for the harvesting yards the best scenarios determined by the environmental criteria do not correspond to the best scenarios identified from the economical point of view: in fact, yards with lower mechanisation level have been considered as more environmentally sustainable

Table 3.11 Weights for sustainability criteria and indicators in order to calculate the global score of each proposed scenario

Criteria and indicators	Weights
Environmental aspects:	
No. of machines used during field operations	0.25
Distance of transport (km)	0.25
Efficiency of the energetic plant (%)	0.50
Total	1.00
Economical aspects:	
Biofuel production cost (€/t)	0.40
No. of storages	0.20
Plant investment cost (k€/kW)	0.20
Plant management	0.20
Total	1.00

Table 3.12 Results obtained applying the MCA to the proposed yards illustrated in Sect. 3.3

Scenarios	Environmental scores	Economical scores
Y1	**A = 1**	B = 2
Y2	B = 2	**A = 1**
Y3	**A = 1**	C = 3
Y4	B = 2	B = 2
Y5	C = 3	B = 2
Y6	B = 2	**A = 1**
Y7	C = 3	**A = 1**
Y8	B = 2	C = 3

The best scenarios are in bold

Table 3.13 Results obtained applying the MCA to the logistics illustrated in Sect. 3.3

Scenarios	Environmental scores	Economical scores
L1	**A = 1**	**A = 1**
L2	**A = 1**	B = 2
L3	B = 2	C = 3
L4	C = 3	C = 3
L5	**A = 1**	B = 2
L6	B = 2	B = 2

The best scenarios are in bold

Table 3.14 Results obtained applying the MCA to the energy plants hypothesised in Sect. 3.3

Scenarios	Environmental scores	Economical scores
E1	B = 2	B = 2
		C = 3
E2	**A = 1**	B = 2
		B = 2
E3	**A = 1**	**A = 1**
		B = 2
E4	C = 3	C = 3
		A = 1
E5	C = 3	C = 3
		A = 1

The best scenarios are in bold

Table 3.15 List of possible agro-energetic chains and results obtained applying the MCA to the proposed scenarios

Chains	Scenarios		Results without weights			Results with weights		
			Environmental scores	Economical scores	Global scores	Environmental scores	Economical scores	Global scores
1	Y1	L1	–	–	–	–	–	–	–
2	Y2	L2	–	–	–	–	–	–	–
3	Y3	L2	E2	1.00	2.25	1.625	1.00	2.40	1.700
4	Y3	L3	E3	1.33	2.25	1.792	1.25	2.40	1.825
5	Y3	L3	E4	2.00	2.50	2.250	2.25	2.60	2.425
6	Y3	L5	E3	1.00	2.00	1.500	1.00	2.20	1.600
7	Y3	L5	E4	1.67	2.25	1.958	2.00	2.40	2.200
8	Y4	L2	E2	1.33	2.00	1.667	1.25	2.00	1.625
9	Y4	L3	E3	1.67	2.00	1.833	1.50	2.00	1.750
10	Y4	L3	E4	2.33	2.25	2.292	2.50	2.20	2.350
11	Y4	L5	E3	1.33	1.75	1.542	1.25	1.80	1.525
12	Y4	L5	E4	2.00	2.00	2.000	2.25	2.00	2.125
13	Y4	L4	E5	2.67	2.25	2.458	2.75	2.20	2.475
14	Y4	L6	E5	2.67	2.00	2.333	2.75	2.00	2.375
15	Y5	L2	E2	1.67	2.00	1.833	1.50	2.00	1.750
16	Y5	L3	E3	2.00	1.75	1.875	1.75	1.80	1.775
17	Y5	L3	E4	2.00	2.25	2.125	1.75	2.20	1.975
18	Y5	L5	E3	1.67	1.75	1.708	1.50	1.80	1.650
19	Y5	L5	E4	2.33	2.00	2.167	2.50	2.00	2.250
20	Y5	L4	E5	3.00	2.25	2.625	3.00	2.20	2.600
21	Y5	L6	E5	3.00	2.00	2.500	3.00	2.00	2.500
22	Y6	L2	E1	1.67	2.00	1.833	1.75	1.80	1.775
23	Y7	L2	E2	1.67	1.75	1.708	1.50	1.60	1.550
24	Y7	L3	E3	2.00	1.75	1.875	1.75	1.60	1.675
25	Y7	L3	E4	2.67	2.00	2.333	2.75	1.80	2.275
26	Y7	L5	E3	1.67	1.50	1.583	1.50	1.40	1.450
27	Y7	L5	E4	2.33	1.75	2.042	2.50	1.60	2.050
28	Y7	L4	E5	3.00	2.00	2.500	3.00	1.80	2.400
29	Y7	L6	E5	3.00	1.75	2.375	3.00	1.60	2.300
30	Y8	L3	E3	1.67	2.25	1.958	1.50	2.40	1.950
31	Y8	L3	E4	2.33	2.50	2.417	2.50	2.60	2.550
32	Y8	L5	E3	1.33	2.00	1.667	1.25	2.20	1.725
33	Y8	L5	E4	2.00	2.25	2.125	2.25	2.40	2.325
34	Y8	L4	E5	2.67	2.50	2.583	2.75	2.60	2.675
35	Y8	L6	E5	2.67	2.25	2.458	2.75	2.40	2.575

but, on the other hand, using a limited number of machines and/or machines characterised by low power, low weight and, in conclusion, low operative capacity, imply higher costs.

On the other hand, logistics which present high environmental sustainability are often also the most convenient solutions from the economical point of view. In fact, minimizing the transport distances significantly reduces the associated environmental pressures (i.e. atmospheric emissions and noise produced by means)

Table 3.16 Six best agro-energetic chains identified through the MCA with or without the application of the weights

Results of the MCA

| Without weights | | With weights | |
Chains	Values	Chains	Values
6 : Y3–L5–E3	1.500	26 : Y7–L5–E3	1.450
11 : Y4–L5–E3	1.542	11 : Y4–L5–E3	1.525
26 : Y7–L5–E3	1.583	23 : Y7–L2–E2	1.550
3 : Y3–L2–E2	1.625	6 : Y3–L5–E3	1.600
8 : Y4–L2–E2	1.667	8 : Y4–L2–E2	1.625
32 : Y8–L5–E3	1.667	18 : Y5–L5–E3	1.650

but usually also the costs, both directly (lower distances) and indirectly (less number of storages and intermediate movements of the biomass).

Finally, concerning the energy plants it is not so easy to establish a correspondence between environmental and economical criteria, because it is necessary to distinguish between thermal and power production: in general, it is possible to detect the higher efficiencies the higher automation level and investment costs.

Table 3.15 shows the possible agro-energetic chains obtained combining all the proposed scenarios and the relative judgements calculated applying the environmental and the economical criteria, either without or with the weights described in Sect. 3.5. The application of the weights changes the group of the best chains selected through the MCA as highlighted for the first six chains in Table 3.16. Particularly, two chains (3 and 32) are identified as profitable applying the MCA without weights, but not if weights are introduced: in this case other two solutions are more interesting, i.e. chains 23 and 18. Moreover, it is possible to analyse that the results do not identify specific yard scenarios but indicate scenarios L2 and L5 for logistics and scenarios E2 and E3, as the most suitable for the fixed criteria. Therefore, on the basis of the results, scenarios that provide only one transport are preferable and direct combustion plants for thermal energy production (small or medium sized) are identified as the most convenient for the specific typology of biofuel.

3.7 Application of the LCA to the Chains Selected Through the MCA

As explained in the previous paragraphs, for the environmental evaluation, it is needed to distinguish between local and global pressures that the agro-energetic chains may originate. For the first ones it is necessary to take into account the principles on which the EIA is based; for the second ones the LCA methodology is needed according to the recent indications of the EC [10].

The LCA methodology has been applied to the chains identified trough the MCA including the weights illustrated in the paragraph for the environmental and the economical criteria, i.e. chains 6, 8, 11, 18, 23 and 26, in order to determine which of these are more suitable to reduce the environmental pressures.

The inventory phase has been conducted as illustrated in Table 3.17 and all the collected data have been implemented in the software GEMIS 4.5 [15].

Concerning the diesel fuel used in the agricultural machines, it has been hypothesised an LHV of 11.86 kWh/kg and a density of 0.8 kg/l; the LHV of wood has been fixed equal to 3.154 kWh/kg. The quantity of diesel fuel required during field operations has been estimated considering an average utilisation of 60% of the nominal power of the machine and a specific fuel consumption of 0.25 kg/kWh [8, 12]. The atmospheric emissions calculated for the yards are referred only to the direct emissions originated by the machines during operations.

Particularly, chain 6 adopts the shredder Nobili of 60 kW with a packaging system; chains 8 and 11 utilise the shredder Berti of 80 kW followed by rural trailer along the field; chain 18 implements a yard with the chipper Pezzolato of 44 kW derived from the forestry sector; chains 23 and 26 hypothesise the use of the small baler Lerda of 50 kW and a next phase of wood processing is carried out by the chipper Pezzolato of 44 kW.

Transports have been modelled considering that local transports within the farm are carried out with a tractor equipped with a rural trailer whilst the other provide the use of highway trucks. Basing on the data of the software GEMIS for the L2 scenario has been considered the process "truck+semi-trailer-D-rural" with a pay load of 25 t and 94.912 gCO_2/t km of emissions; whilst the L5 scenario corresponds to the process "truck-highway-EURO 4-20-28 t" with 132.64 gCO_2/t km of emissions according to the European emission standard EURO 4.

For the energy conversion phase each hypothesised plant has been characterised by the operation time, the life time, the power, the efficiency, the electricity required by the plant and the main construction materials. In addition, the energetic plants are equipped by a multicyclone to depurate the flue gases assuring a reduction of about 92% of particulate.

The obtained results are reported in the followings. Particularly, chain 18 is the most suitable because it is able to minimize the impacts analysed trough the LCA, i.e. the CO_2eq and the CER (see Chap. 1). However, the MCA has identified chain 18 only as the sixth between the most appreciable ones, also if only the environmental criteria are taken into account without combining them with the economical ones. This is only an apparent contrast because actually the criteria introduced trough the LCA are not taken into account in the MCA and therefore must be evaluated as additional (Figs. 3.8, 3.9).

Considering that the contributions to the CO_2 equivalent emissions (and also to the CER) of logistics and energetic conversion scenarios are about constant from chain to chain, it is obvious that the harvesting and processing phases can make the

Table 3.17 Inventory data for the most sustainable chains according to the MCA

Chain	6	8	11	18	23	26
Harvesting scenarios	Y3	Y4	Y4	Y5	Y7	Y7
Operative capacity (t/h)	0.6	0.9	0.9	2.0	3.0 (baler) 2.0 (chipper)	3.0 (baler) 2.0 (chipper)
Fuel consumption (kg/h)	9	12	12	6.6	7.5 (baler) 6.6 (chipper)	7.5 (baler) 6.6 (chipper)
Operation time (h/year)	1.67	1.11	1.11	0.50	0.83	0.83
Life time (year)	20	20	20	20	20	20
Fuel quantity (kWh/kWh)	0.0564	0.0501	0.0501	0.0124	0.0218	0.0218
CO_2-eq emissions (g/MWh)	14,910	13,253	13,253	3,280	5,765	5,765
Logistic scenario	L5	L2	L5	L5	L2	L5
Means typology	Highway truck	Tractor + rural trailer	Highway truck	Highway truck	Tractor + rural trailer	Highway truck
Transport distance (km)	40	15	40	40	15	40
Energetic scenario	E3	E2	E3	E3	E2	E3
Operation time (h/year)	2,200	2,200	2,200	2,200	2,200	2,200
Life time (year)	15	10	15	15	10	15
Power (MW)	0.8	0.1	0.8	0.8	0.1	0.8
Efficiency (%)	85	85	85	85	85	85
Electricity from grid (kWh/kWh)	0.02	0.02	0.02	0.02	0.02	0.02
Metal/Steel (kg/MW)	0.50	0.50	0.50	0.50	0.50	0.50

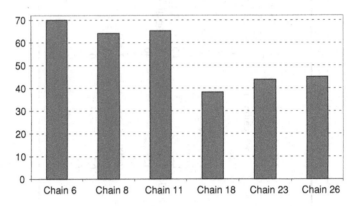

Fig. 3.8 CO₂eq emissions for the most suitable chains identified trough the MCA

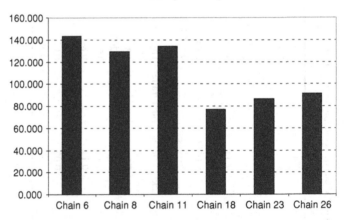

Fig. 3.9 CER for the most suitable chains identified trough the MCA

Table 3.18 CO₂eq emissions for the most suitable chains identified through the MCA

		Chain 6	Chain 8	Chain 11	Chain 18	Chain 23	Chain 26
Total	gCO₂/kWh	69.874	64.129	65.359	38.267	43.794	45.025
Y scenario	gCO₂/kWh	Y3 41	Y4 36	Y4 36	Y5 9	Y7 16	Y7 16
L scenario	gCO₂/kWh	L5 2	L2 1	L5 2	L5 2	L2 1	L5 2
E scenario	gCO₂/kWh	E3 27.4	E2 27.6	E3 27.4	E3 27.4	E2 27.6	E3 27.4

Table 3.19 CER for the most suitable chains identified through the MCA

	Chain 6	Chain 8	Chain 11	Chain 18	Chain 23	Chain 26
W/kWh	*143*	*129*	*134*	*77*	*87*	*91*

difference on the total amounts. Particularly, it is possible to highlight (see Tables 3.18 and 3.19) those chains where the Y scenarios results as less pollutant are the best: this is the case of chains 18, 23 and 26 in comparison with the solutions 6, 8 and 11 which present values of indicators also more than four times higher.

References

1. Ambiente Italia (2002) Programma di azioni a supporto dell'iniziativa delle amministrazioni locali in attuazione della convenzione quadro sui cambiamenti climatici. Ambiente Italia, Roma
2. ANPA (2001) I rifiuti del comparto agroalimentare. ANPA-Unità Normativa Tecnica, Roma
3. Area Science Park (2006) Energia dalle biomasse: le tecnologie, i vantaggi per i processi produttivi, i valori economici e ambientali. Consorzio per l'AREA di ricerca scientifica e tecnologica di Trieste AREA Science Park, Trieste
4. ARSIA (2004) Le colture dedicate ad uso energetico, Quaderno n.6. ARSIA, Firenze
5. Beaufoy G (2001) The environmental impact of olive oil production in the European Union: practical options for improving the environmental impact. European Forum on Nature Conservation and Pastoralism and the Asociacion para el Analisis y Reforma de la Politica Agro-rural, Brussels
6. BIOSIT (2003) GIS based methodology and algorithm. BIOSIT project, LIFE00 ENV/IT/000054, Firenze
7. Chechi G (2007) Recuperi energetici in azienda agricola con indirizzo olivicolo e fruttifero. Master thesis in agro-engineering at the University of Florence, A.A. 2006/2007
8. Cini E, Recchia L (2008) Energia da biomassa un'opportunità per le aziende agricole. Pacini Editore, Pisa
9. CTI (2003) Biocombustibili: specifiche e classificazione, Raccomandazione R03/1. Milano
10. EC (2010) Report from the Commission to the Council and the European Parliament on sustainability requirements for the use of solid and gaseous biomass sources in electricity, heating and cooling, SEC 65-66 (2010). Brussels
11. EEA (2007) Estimating the environmentally compatible bioenergy potential from agriculture, Technical report. EEA, Copenhagen. ISBN 978-92-9167-969-0
12. ENAMA (2005) Prontuario dei consumi di carburante per l'impiego agevolato in agricoltura. ENAMA, Roma
13. ENEA (2005) Le fonti rinnovabili: lo sviluppo delle rinnovabili in Italia tra necessità e opportunità. ENEA, Roma
14. ENEA (2009) Usi termici delle fonti rinnovabili. In: Proceedings of the workshop "Usi termici delle fonti rinnovabili", Roma, 11 novembre 2009
15. GEMIS (2010) http://www.oeko.de/service/gemis
16. Ghafghazi S, Sowlati T, Sokhansanj S, Melin S (2009) A multicriteria approach to evaluate district heating system options. Appl Energy. doi:10.1016/j.apenergy.2009.06.021
17. ITABIA (2003) Rapporto sullo stato della bioenergia in Italia al 2002. Roma
18. Recchia L, Daou M, Rimediotti M, Cini E, Vieri M (2009) New shredding machine for recycling pruning residuals. Biomass Bioenergy 33:149–154

19. Recchia L, Cini E, Corsi S (2010) Multicriteria analysis to evaluate the energetic reuse of riparian vegetation. Appl Energy 87:310–319
20. Riva G, Alberti M (2005) Il controllo e la certificazione delle emissioni in impianti a legna. Ambientediritto.it, Messina
21. Spinelli R, Picchi G (2010) Industrial harvesting of olive tree pruning residue for energy biomass. Bioresour Technol 101:730–735
22. Woodland Energy (2009) La filiera Legno-Energia come strumento di valorizzazione delle biomasse legnose agroforestali, Woodland energy project, Probio-MiPAAF programme. LCD srl, Firenze

Chapter 4
Agricultural and Forestry Mechanization

4.1 Introduction

In recent years, reducing the environmental impact has become an increasingly important subject for the general public. The objectives imposed by the Kyoto Protocol regarding the reduction of greenhouse gas emissions have guided many research projects towards the supply of alternative "clean" energy that does not come from fossil sources.

In Europe, "20-20-20" programmatic objectives have been set for 2020, i.e. 20% reduction in greenhouse gas production, 20% increase in the use of alternative energy sources in our final energy consumption and, finally, 20% increase in energy efficiency for transformation processes.

Within the context of renewable energy, over the next few years the wood biomass sector predicts that there will be a broad potential for development due to the abundance of raw materials available as well as the simplicity of using the same. In Italy, currently the main sources of wood biomass are from woodland management, farming residue, dedicated crops and, finally, waste matter from agro-industrial processes.

According to the last 2009 biomass energy report [7] Italy has a total installed biomass power capacity of 7,558 MWt, contributing 5.2 Mtep to our country's primary energy production, which corresponds to around 2.7% of our total energy requirements, putting us in the fifth place amongst European countries for the production of energy from agro-forestry biomasses.

The most widespread and commonly used biomasses are: firewood, which is generally used in small stoves for residential use with low output, for the direct production of heat or to supplement the heating system of individual residential units or buildings consisting of several units; chips, used in high conversion efficiency boilers with maximum power capacity of 1 MWt used in a similar way and, finally, pellet stoves, over 1 million of which were installed in Italy in 2009, making our country the top European nation for this type of installation [7].

L. Recchia et al., *Multicriteria Analysis and LCA Techniques*,
Green Energy and Technology, DOI: 10.1007/978-0-85729-704-4_4,
© Springer-Verlag London Limited 2011

In Italy, in recent years, thanks to public support, the development of short local-district energy chains has received a considerable boost leading to the decentralization of energy production, the downscaling of installations and, thus, a wider use of waste from agricultural, forestry and agro-industrial production and the expansion of crops dedicated to the production of chips.

The biggest problem in the use of wood biomass is the difficulty of exploiting the same in an economically viable way. Some difficulties are due to the rather patchwork territorial distribution and seasonality of the supply which makes intermediate storage necessary before and after the consignment of the material to the energy production plant.

Moreover, wood materials have an energy density that is 2/4 times lower than that of fossil sources and so, in order to produce the same amount of energy, huge volumes need to be mobilized requiring the realization of chains that must be carefully organized in order to be economically viable.

This is necessary because each wood-energy chain involves multiple phases (supply, primary transformation, transportation, intermediate and final storage, conversion to solid bio-fuel, delivery to plants, energy production) and, as a result, many potential problems can be triggered during the process. For this reason, the chain has to be implemented with an integrated process approach since the badly structured planning of the chain could mean that the final cost for the energy produced is not competitive with respect to traditional fossil fuels.

Moreover, an efficient utilization of biomasses requires a meticulous organization of the yard in order to guarantee maximum productivity. At the same time, the size of the yard should be relative to the specific working conditions applied.

Too often one aspect that is not carefully assessed when realizing a wood–energy chain is the environmental sustainability of the processes set up. The implementation of a chain for the exploitation of energy from wood biomasses should not disregard the concurrent environmental and economic sustainability of all the production phases.

With regard to organizing the chain and planning the processing systems, the possibility of using local biomasses for energy purposes is a good way of reducing CO_2 emissions and guaranteeing the sustainable development of energy supplies in different production contexts.

Alongside traditional supplies of residual biomasses, part of the studies on alternative bio-energy have concentrated on the production of energy from wood obtained from arboreal species with short coppicing cycles; this is the case for short and medium rotation coppice [32].

These crops, managed using different planting and cultivation models, have in common the use of rapid growth arboreal species with strong shoot regrowth after cutting such as willows *Salix* ssp., poplars *Populus* spp., False Acacias *Robinia pseudoacacia* and Eucalyptus *Eucalyptus regnans*. These differ in terms of the species as well as the lifetime which ranges from 8 to 12 years and 2 to 3 years for short cycle species and up to 15 years for medium rotation, with cycles varying from 4 to 6 years, as well as planting density, 8,000–1,000 plants per hectare for short rotation cycles and 1,000–1,800 plants per hectare for medium rotation coppice.

The final product obtained is wood chips which are subsequently used for energy conversion for the production of heat and/or electricity using thermo-chemical (direct combustion, pyrolis and gasification) or biochemical processes (anaerobic digestion).

4.2 Goals, Definitions and Decision Makers' Typology

This work aims to present the multicriteria approach as a preliminary decision-making tool for studies aimed at determining the environmental impact of certain wood–energy chains by assessing their life cycle (LCA: Life Cycle Assessment).

The paper provides some indications regarding the method of application of the multicriteria approach in deciding upon the suitability of the reference criteria (crops, harvest, logistics) in relation to the environmental and economic aspects [35]. The first case study analysed concerns chains for the production of wood biomasses from dedicated tree crops with different coppicing rotation cycles. The second part of the treatise includes an example of the application of the methodology in wood–energy chains for the production of energy from wood biomass obtained from woodland areas.

The text includes some examples of crop scenarios, harvesting techniques, production diagrams and diagrams displaying normal energy use for the ordinary management of farming and woodland businesses, in order to identify the best scenarios from the point of view of environmental and economic sustainability according to the decisional criteria adopted [11, 12, 35].

The hypothetical scenarios were studied using normal farming and management practices as a reference, making the necessary adjustments. The management methods, materials and machinery used are given for each scenario.

Each chain has been classified using the assessments determined by the decisional criteria based on the environmental and economic aspects. The environmental impact of the best scenarios identified according to the multicriteria analysis was subsequently measured by means of a LCA.

4.3 Definition of SRC and MRC Scenarios

The cultivation of coppices for biomass production presents further complexities with respect to chains featuring processes aimed exclusively at obtaining a secondary product derived from a principal activity, because variables related to the production of the raw material also come into play. It is not, therefore, sufficient to simply identify the best technology for the harvesting phase or logistics, but it is necessary to choose the best production site for the most suitable variety, as well as the right cultivation and management model on which the subsequent harvesting, transformation, logistics and energy transformation phases will depend.

Moreover, the success of Short and Medium Rotation Coppice, with respect to other methods of cultivation, depends on the correct application of modern farming practices (weed killing, mechanical containment of weeds, fertilizing, irrigation, etc.), whilst taking into account the economic and environmental sustainability.

The myriad choices that can be implemented for wood–energy chains using coppices for biomass make the definition of more advantageous solutions quite complex. The multicriteria analysis thus becomes an important support tool in the decision-making process that can help to clarify situations in which numerous possibilities are available. The study process has been built on a series of comparative criteria which permit the identification of the best scenarios from the point of view of environmental and economic impact. The scenarios differ in terms of cultivation techniques and intensity, i.e. the number and type of cultivation interventions, type of harvesting yard used and, finally the type of logistics.

4.3.1 Description of Yards and Crop Management

With regard to cultivation, four different planting solutions of poplar for biomass were examined, defined according to normal ordinary management practices used both nationally and internationally. In particular, all the phases were analysed, from planting to the final extirpation phase see Table 4.1. The main data studied are given in Table 4.2.

The four production contexts (named Y1L Y2H Y3L Y4H where "L" means *low input* and "H" *high input*) were created using as a classification variable the energy input requirement in terms of use of non-renewable fossil materials as well as input of synthetic chemical products.

The planting layouts studied have the following characteristics:

Y1L, Y2H	Y3L, Y4H
• Planting density 8000 cuttings/ha	• Planting density 1667 rods/ha
• Crop lifetime 10 years	• Crop lifetime 15 years
• Rotation length 2 years	• Rotation length 5 years
• Planting layout 2.5 m × 0.5 m	• Planting layout 3 m × 2 m
• Average biomass yield 12 t_{db} ha/year	• Average biomass yield 14 t_{db} ha/year

The preparatory phases for planting included two management methods also found in other Italian experiments [25, 38]. The first, implemented in low input scenarios, involved 0.5 m plowing, after which the soil was worked with a power harrow. The second, for high input scenarios, consisted of crossed double rooting at a depth of 0.7 m followed by the same phases used for low input scenarios.

For transplanting cuttings, the operational capacities obtained in yards located in central-northern Italy were used as a reference [4]. In particular, for SRC tree

Table 4.1 Crop management scenarios used for Y1L Y2H Y3L Y4H

Crop management Y1L	Year									
	1	2	3	4	5	6	7	8	9	10
Ploughing	x									
Deep fertilizing	x									
Power harrowing	x									
Planting	x									
Pre-emergence chemical herbicide	x									
Ridging	x									
Harrowing	xx	xx	x	xx	x	xx	x	xx	x	xx
Organic fertilization			x		x		x		x	
Treatments	x	x	x	x	x	x	x	x	x	x
Harvesting		x		x		x		x		x
Uprooting										x

Crop management Y2H	Year									
	1	2	3	4	5	6	7	8	9	10
Subsoiling	x									
Ploughing	x									
Deep fertilizing	x									
Power harrowing	x									
Planting	x									
Pre-emergence chemical herbicide	x									
Ridging	x									
Harrowing	xx	x	x	x	x	x	x	x	x	x
Mulching		x	x	x	x	x	x	x	x	x
Mineral fertilizing			x		x		x		x	
Irrigation	x		x		x		x		x	
Treatments	xx	xx	xx	xx	xx	xx	xx	xx	xx	xx
Harvesting		x		x		x		x		x
Uprooting										x

Crop management Y3L	Year														
	1	2	3	4	5	6	7	8	9	10	11	12	13	14	15
Ploughing	x														
Deep fertilizing	x														
Power harrowing	x														
Planting	x														
Pre-emergence chemical herbicide	x														
Harrowing	xx	x				xx	x				xx	x			
Mulching					x					x					x
Mineral fertilizing		x			x					x					
Treatments		x	x	x		x		x		x	x	x	x		
Harvesting					x					x					x
Uprooting															x

(*continued*)

Table 4.1 (continued)

Crop management Y4H	Year 1	2	3	4	5	6	7	8	9	10	11	12	13	14	15
Subsoiling	x														
Ploughing	x														
Deep fertilizing	x														
Power harrowing	x														
Planting	x														
Pre-emergence chemical herbicide	x														
Harrowing	xx				x	xx				x	xx				x
Mulching				x						x					
Mineral fertilizing	x					x					x				
Irrigations	x	x				x					x				
Treatments	x	x	x	x		x	x	x	x		x	x	x	x	
Harvesting				x						x					x
Uprooting															x

farms, reference was made to a two-row transplanter with an operational capacity of 0.6 ha/h, and for MRC tree farms, a stripling transplanter with a working capacity of 0.4 ha/h.

During the planting management phases, the distinction between low and high input, respectively in the Y1L Y3L and Y2H Y3H scenarios, referred to the higher number of mechanical and chemical interventions, the type of fertilization adopted, whether organic or chemical, as well as the number and power of the machines used.

For dimensioning the productivity of the fertilizing operations, only the costs of the technical distribution methods were considered given the extreme variability of the composition of the same and, consequently, the market prices on a national level, making it difficult to identify a representative price per ton of product. For scenarios that involved the use of synthetic fertilizers, the cost of distribution machinery as well as the raw materials was calculated, referring to a fertilizer with a titer of 8:24:24 and a market price of 0.43 €/kg determined by the prices on the main Italian markets [10].

Pest control in the four cultivation models differed in terms of the number of interventions per year but not the type of pest control product used. For treatments against the main trunk-boring and xylophagous insects (e.g. *Cossus cossus*, *Cryptorhynchus lapathi*, etc.) or defoliators (e.g. *Chrysomela populi*, *Leucoma salicis*, etc.), the standard treatment was used as a reference for all the scenarios, with a volume of 600 l of water per hectare to which 0.84 l/ha of Fenitrothion p.a., 0.36 l/ha of Chlorpyrifos p.a. and 0.06 l/ha of Cypermethrin p.a. were added.

Table 4.2 Main features of the means and materials used in the scenarios Y1L Y2H Y3L Y4H

Crop management		Machinery	Diesel consumption (kg/ha)	Materials	Required time (h/ha)
Subsoiling		Tractor 210 kW + subsoiler	126	–	3.17
Plowing		Tractor 150 kW + three-furrow plough	56.67	–	2.12
Power harrowing		Tractor 90 kW + disc harrow	20.77	–	1.28
Cow dung deep fertilizing		Tractor 80 kW + Manure spreader capacity 10 m^3	6.55	Cow dung 50,000 kg	0.91
Mineral deep fertilizing		Tractor 65 kW + centrifugal spreader	1.53	NPK kg/ha 150/120/250	0.26
Planting cuttings-rods	SRC	Tractor 65 kW + two rows transplanter	14.63	8,000 cuttings	1.23
	MRC	Tractor 65 kW + single rows transplanter	25.59	1,667 rods	2.50
Ridging		Tractor 65 kW + ridging wing	5.08	–	0.69
Irrigation		Tractor 45 kW + pivot	21.26	40 mm H_2O[a]	6
Harrowing	SRC	Tractor 90 kW + disc harrow 2,5 m	13.89	–	1.14
	MRC	Tractor 90 kW + disc harrow 3 m	13.50	–	0.95
Mulching	SRC	Tractor 80 kW + grass mulcher 2,5 m	14.40	–	1.33
	MRC	Tractor 80 kW + grass mulcher 3 m	12	–	1.11
Top dressing		Tractor 65 kW + centrifugal spreader	1.53	Fertilizer NPK 8:24:24[b]	0.26
Fertilizing		Tractor 120 kW + liquid manure spreader tank	12	Liquid manure fertiliser 20,000 kg/ha	1.11
Chemical weed control		Tractor 65 kW + boom sprayer	4.25	Trifluralin + linuron + alaclor 0.8 + 0.4 + 1.4 kg/ha + 400 l H_2O/ha	0.73
Treatments		Tractor 65 kW + cannon sprayer	2.09	Fenitrothion + chlorpyrifos + cypermethirin 0.84 + 0.36 + 0.06 kg/ha + 600 l H_2O/ha	0.29
Uprooting	SRC	Tractor 100 kW + stump cutter	56.47	–	3.14
	MRC	Tractor 100 kW + stump cutter	47.06	–	2.61

a Quantity of water distributed for each intervention

b Fertilizer quantity distributed for each intervention 600 kg/ha Y2H, 800 kg/ha Y3L, 1,000 kg/ha Y4H

With regard to fungal parasites, since clones which are resistant to the most frequent pathologies are normally used (*Cytospora* spp., *Discosporium populeum*, etc.), no specific treatments were necessary [2, 18].

The cost per single treatment was calculated by adding together the purchase cost of the active ingredients determined by referring to the market prices of the commercial formulations, and the cost of distribution carried out by a cannon sprayer with a range of 14–16 m coupled with a 65 kW tractor.

The soil management and pest control operations were assessed based on normal practice using disk harrows, choppers and harrows with rotating devices operated by the *power take off* (pto). The calculation of fuel consumption and operating costs referred to operating machinery coupled with 80–90 kW tractors.

Chemical weed control during planting involved the use of residual herbicides acting in germination with 30–40 days' cover using a mixture of Triflura-lin + Linuron p.a. 0.8 + 0.4 l/ha against dicotyledon species to which 1.4 l/ha p.a. of Alachlor was added for graminaceous species, all of which was distributed by machines working simultaneously on several rows, with average operating capacities of 2 ha/h for both plantation types.

In the Y2H and Y4H scenarios, emergency irrigation interventions were planned for the years following coppicing to promote vegetative recovery, to be carried out in the spring–summer season, distributing 400 mm of water by means of a pivot irrigator, the pump of which was operated by a tractor with nominal power of 45 kW.

For each cultivation scenario, the final extirpation phase of the plantation was carried out by a milling machine operated by a 100 kW tractor working at a speed of 1–1.5 km/h.

The cost of the cultivation operations was calculated by comparing the values obtained through direct determination with the actual costs obtained from Italian experiments and the prices of mechanical–agricultural work in the regions of Tuscany, Emilia-Romagna, Veneto and Lombardy [3, 15, 29], subsequently adjusted to ideal working conditions.

Based on the considerations and premises described, a production cost per ton of product was defined for the various scenarios, varying between 58 and 112 €/t_{dm}, not including harvesting costs and public incentives that vary considerably throughout Italy. The author's state that the costs identified do not reflect the economical hypotheses of all the production contexts, therefore they invite the reader to consider their estimate as just one of the possible estimates.

4.3.2 Description of Harvest Scenarios

With regard to the harvesting scenarios, yards normally used for SRC and MRC harvests were studied. The characteristics of poplar for biomass and, more in general, crops for biomass production from tree farms, including the consistence of the wood and contemporary emission of several shoots, made it necessary to use

Table 4.3 Main features of the harvesting yards scenarios

Scenario	Yard mechanisation	Power requirement (kW)	Harvest yard investment (k€)	Hourly production (t/h)
H1	Cut and chip loaders Three tractors Two rural trailers	160 + 95 + 95 **350** (300–400)	**270** (250–300)	20–25
H2	Cut and chip harvester Two tractor Two rural trailers	340 + 95 + 95 **530** (450–550)	**620** (550–650)	35–45
H3	Excavator based harvester extraction agricultural tractor with trailer and crane mobile chipper 150 kW	75 + 70 + 150 **295** (250–350)	**280** (250–350)	8–12
H4	Harvester Skidder mobile chipper 300 kW	140 + 65 + 300 **515** (500–600)	**570** (550–600)	18–20

() Range of possible values

high power machines for cutting and shredding. These differed in terms of the level of mechanization, purchase investment necessary, working capacity and versatility of use.

For each of the four yards identified as H1,H2,H3,H4, the following was determined: the productivity of the yard, the total kilowatts of the machinery, the cost per hectare and the necessary investment. The costs per hectare were calculated by comparing and revising the references available from the relevant literature with data from experiments carried out in Tuscany [8], making the necessary adjustments and corrections. The investment necessary for purchasing machinery was deemed to be the average value obtained from the price lists available on the web and from dedicated scientific magazines [20–22]. Various references were used to define the hourly productivity values comparing them with the most significant experiments carried out in Italy [1, 2, 6, 13, 23, 33]. Table 4.3 summarizes the study yards proposed.

Yard H1, used in SRC single or double row tree farms, required the use of a 150–180 kW tractor equipped with a head for harvesting and chipping alongside two 95 kW tractors towing 15–20 m^3 dump trailers for transporting the chips to the storage areas Fig. 4.1. The operating machine could be tractor-mounted or semi-mounted coupled to the rear hydraulic three point hitch or, where power permitted, to tractors with a hitch and front power take-off. The base units of this machinery consisted of a cutting system with two circular toothed blades rotating in the opposite direction that cut the tree at around 80–150 mm from the ground in order not to compromise the stumps, and conveyor systems with various configurations such as auger conveyors, feeder rollers and shaft conveyors; these directed the trees felled towards a shredding system consisting of a rotating cutting element that transformed them into chips. Finally, there was an automatic or manual side

Fig. 4.1 Yards for the "in continuous harvest" of SRC. On the left H1 yard standard tractor-mounted SRC header harvester; on the right cut and chip harvester on willow (from http http://www.claas.com (2010))

banding outlet pipe for the chips produced, to direct them towards a harvester trailer. The hourly productivity rate of this type of harvesting yard is around 20–25 t_{wb}/h of chips [2, 26, 39].

In yard H2 the harvest was carried out using machinery originally designed for harvesting ensiling maize and, more in general, forage crops Fig. 4.1. It is, in fact one of the cut and chip loaders better known as a forage harvester to which modifications have been made to the harvester head and cutting elements so that it can be used for harvesting trees. These machines can be used on trees with an average base diameter of 80–100 mm. Currently, there are no technologies that permit the continuous harvesting–shredding of larger sized trees. The machine basically consisted of two parts:

- The drive unit that carries out the functions of movement, power distribution and wood shredding;
- The cutting unit known as the "head" that cuts and conveys the trees to the shredder.

After shredding, the product was directly discharged onto a trailer using a side banding chute. During the operational phase, the forage harvester was accompanied by two 95 kW tractors towing 20 m³ dump trailers for transporting the chips. Yard H2 was similar to yard H1 in terms of logistics and work plan, although differed with regard to the high level of productivity that could be achieved, on average from 30 to 50 t_{wb}/h, the weights (8–12 t) and the fuel consumption (35–45 vs. 20–25 l/h) that were much higher, as well as the lower operational flexibility and investment necessary of around 500,000 € [36, 39].

Currently, yards H1 and H2 are the only two mechanization typologies that can be used for harvesting in short rotation coppice. Although myriad configurations exist, they all follow the operating principles and work plans of the machinery described above.

To date, it has been possible to manage medium rotation coppice harvesting with several levels of mechanization compatibly with the technologies available.

This is possible due to the longer coppicing rotation that allows the trees to grow to such a size as to make the use of different types of mechanization possible. To date, forestry harvesting equipment is the option that can best be adapted to these forms of cultivation. The perfect geometry of the tree farm, the position of the land which is usually on a plain or with limited sloping, makes it easier to use equipment such as forestry processors, harvesters, feller bunchers, forwarders and skidders. The many studies carried out on agricultural–forestry mechanization and the experiments conducted worldwide confirm the versatility and high productivity of yards using these methods even in areas which are difficult to access [1, 32].

Their use in MRC harvesting is made possible not just by the versatility and productivity of this machinery but also the assortments that can be obtained from this type of tree farm which, with a coppicing rotation of 4–6 years, can result in a differentiation of the products and, consequently, the efficient use of machinery designed for the forestry sector in a much more advantageous and accessible environment.

The H3 and H4 yards, where the tree was cut, the whole plant extracted and chipping subsequently carried out in dedicated areas near the plantation, differed in terms of the cutting-extraction technology used and the productivity of the chipping unit.

In the H3 yard, the cutting was carried out by a 75 kW, 12 t weight hydraulic excavator equipped with a harvesting and processing head or "processor". Today the hydraulic excavator represents an important resource for the agricultural and forestry sector mainly due to the characteristics of the machine itself which include easy operation and multiple processors, making it a tool that can mechanize operations in various production contexts. The versatility of the excavator is also a result of the high performance of the hydraulic system that can satisfy the demands of even the most complex equipment, as well as the broad range of the motor–cabin–arm unit that makes it possible to work easily and safely even in difficult conditions.

The harvesting head mounted in place of the bucket carried out the cutting function by means of a device consisting of a grooved bar with a cutting chain running along the same. During the working phase, the tree was attached using three or four hydraulically controlled metal arms; the guide bar then came out of the base of the processor and did the cutting.

The processor can be equipped with rollers and knives that, in addition to delimbing and cutting, also simultaneously debark the tree so that this operation does not have to be carried out at the sawmill. Other advantages include better harvesting yard operating quality since it is possible to direct the fall of the trees, thus reducing the risk of damage to the remaining plants. The wood is also stacked in an organized way at the side of the working area for MRC tree farms or, in the wood or landing, in forestry areas.

The "tree farm extraction" phase followed which was carried out by a 75 kW tractor with a two powered axle trailer and a working load of up to 10 t, as well as a crane with a hydraulic grapple that takes the whole tree and places it in the trailer, subsequently unloading it near the chipper. This last phase was carried out by a chipper with an average power of 150 kW placed on a dual axis trailer with a crane that can work with trees with a base diameter of 0.4 m, which has an average hourly productivity of 8–12 t [1, 24, 32].

Equipment with the highest level of mechanization was used in yard H4. Cutting was carried out by a harvester machine designed for cutting processing functions in the forestry sector. It consisted of a drive unit with articulated frame and 4 or 6 driving wheels, a 9–11 m articulated hydraulic arm crane at the end of which a combined head was mounted, controlled by a joystick. These machines were equipped with sensors and software systems for easy operation and spacial monitoring of all the cutting variables. The operator has to pre-set all the parameters required for the commercial assortment requested using a graphic interface on the monitor mounted in the cabin, and also identify and hitch the tree to be felled.

The cut trees were removed using a skidder (modified forestry tractor) with a simple grapple. This machine, which is similar to a farm tractor, differs in terms of the high level of robustness and articulation offered by the reinforced frames which have a central steering joint that allows them to easily cope with the uneven woodland conditions, thanks to the isodiametric wheels with increased section that simultaneously increase the surface adhesion and reduce the damage to the wood's subsoil. Generally they have one or two winches or large extraction grapples. The working phase involved using the hydraulic grapple to simultaneously hitch several felled trees which were then taken to the chipper.

Finally, chipping was carried out by a 350 kW chipper placed on a truck that can work wood with a maximum diameter of 0.65 m, with average hourly productivity of 10–18 t [1, 24, 32].

In yards H3 and H4, the higher hourly productivity of the chipper with respect to the other harvesting yard machinery, made it necessary to determine the values by presuming optimal theoretical working conditions, i.e. when the chipper has enough material available to avoid having to halt operations.

4.3.3 Description of Logistics Scenarios

The final scenario definition phase consisted of identifying the logistics options for transporting the chips to the energy conversion plant. Five logistical scenarios were defined, respectively L1, L2, L3, L4, L5, that differed in terms of the distance covered, the number of storage sites for processed materials, the carrying capacity and the power of the equipment used Table 4.4.

The distances covered varied from 1 km to a maximum of 100 km, thus representing the simplest situations, where the supply and consumption of chips is local-district, and the more complex situations that require dedicated storage centers, long distance transport vehicles and large conversion plants.

The vehicles used for the five logistical scenarios were combinations of the following units:

Table 4.4 Main features of the hypothesized logistics scenarios for the wood chips transport

Scenario	First Transport	First Storage	Second Transport	Second Storage	Carrier capacity (m³)	Transport power requirement
L1	1	Farm storage	–	–	Rural trailer	**40** (20 + 20)	190
L2	5	Dedicated Areas	30	Plant storage	Rural trailer + truck trailer	**115** (20 + 20 + 75)	190 + 320
L3	100	–	–	–	Truck trailer with crane	**75**	320
L4	80	Dedicated Areas	35	Plant storage	Truck trailer with crane + truck trailer	**150** (75 + 75)	320 + 320
L5	10	Farm storage	5	Plant storage	Rural trailer + rural trailer	**80** (20 + 20 + 20 + 20)	190 + 190

Table 4.5 Definition of evaluation criteria for SRC and MRC scenarios

Criteria	Weights		
	A	B	C
Environmental Aspects			
1. kW management phase	≤250	>250 and ≤350	>350
2. No. of chemical pest and weeds control treatment	≤10	>10 and ≤15	>15
3. No. of irrigation	–	> 0 and ≤3	>3
4. Type of fertilization	Organic	Mixed	Chemical
5. Distance km	≤15	>15 and ≤50	>50
6. Carrier capacity m³	>100	>50 and ≤100	≤50
7. kW transport	≤350	>350 and ≤550	>550
8. Hourly productivity t/h	≤15	>15 and ≤30	>30
9. kW harvesting yard	≤250	>250 and ≤450	> 450
Economic Aspects			
10. Production cost €/year	≤850	>850 and ≤1,000	>1,000
11. No. of storage	–	1	2
12. Harvesting yard investment k€	≤300	>300 and ≤500	>500
13. Harvesting cost €/ha	≤500	> 500 and ≤700	>700

- Tractor with farm trailer with a carrying capacity of 20 m³;
- Truck and trailer with a total carrying capacity of 75 m³;
- Truck with crane + hydraulic grapple coupled with a trailer that has a total carrying capacity of 75 m³.

With regard to environmental and economic sustainability, the logistical organization had to be structured using machines and technologies that minimize costs and impact. For this purpose, single passage solutions were useful as they guarantee rapid loading-transfer of the chips between the harvesting unit and the transport unit. This would also avoid the qualitative decline of the biomass due to contamination of the chips with impurities such as soil and stones which are often found in multiple phase systems.

However, whereas on the one hand single passage systems improve certain logistical aspects, on the other hand a less "valuable" biofuel is obtained due to the high water content which, in some cases, is more than 50% of the fresh substance, leading to the subtraction of heat during the combustion phase that results in lower energy yields. Moreover, the higher the water content of the wood, the greater the specific weight and, consequently, the higher the transportation costs.

Further important considerations that are also valid for the logistics of transporting biomasses from dedicated crops are shown in Sect. 4.8.2.

The storage areas identified were those normally found in ordinary practice, i.e. at the farm, in dedicated areas next to the biomass plant or inside dedicated structures near the energy transformation plant.

It was difficult to eliminate the storage phase due to the different timing between supply and utilization, resulting in additional costs for the management of large volumes of biomass since the latter is characterized by a relatively high weight and volume per unit of energy produced.

4.4 Decision Criteria Identification

As already explained in Sect. 4.1, in order to make an evaluation it is necessary to define the decisional criteria. This phase represents an important stage in the entire analysis process and, for this reason, should be set up using significant comparative variables in such a number as to make the comparison as representative as possible. In this case study, the criteria identified referred to environmental and economic aspects. Within these two macro-categories, they were further divided according to criteria that concern aspects linked to cultivation techniques at the harvesting yards and, finally, the logistics for transporting the chips Table 4.5.

The decisional criteria related to cultivation techniques were drawn up by adopting, as a discriminant, the energy input requirement in terms of use of non-renewable fossil materials as well as the input of synthetic chemicals. As a result, the preferred or "best" scenarios were those in which: low power mechanical machinery was used,[1] the number of chemical interventions for pest and weed control were limited, and fewer irrigation interventions and organic fertilizers were adopted.

With regard to the environmental aspects, similar considerations to those used for the cultivation techniques were applied to define the harvesting methods, i.e. yards with high hourly machinery productivity rates were preferred since in that unit of time the ratio[2] between kilogram of fuel used and tons of chips produced was lower, as well as yards using machinery with a lower power requirement.

From the point of view of logistics, the optimal chains were those in which transportation distances were limited, that used machinery combinations with a higher carrying capacity and reduced power demand. The use of machinery with a high volumetric load during all phases undoubtedly made it possible to optimize the logistics because often, when transporting chips, it is impossible to fully exploit the maximum carrying limit of the units used as the volumetric load limit is reached first.

The economic aspects refer to the production cost of the biomass in terms of the price of the trees standing, the number of chip storage yards, the investment necessary to purchase harvesting technologies and the harvesting cost per hectare for each specific yard.

---

[1] Generally speaking it can be stated that the increase in kilowatt supplied by the machinery corresponded to an increase in fuel consumption within the time unit. This was determined by the following formula: $G = c_s\ dP$ (in kilogram diesel oil g/h) where c_s specific consumption typical of an agricultural diesel engine (kg/kWh) at maximum power, d the maximum power quota requested, P the power of the engine of the driving or self-propelled machine used.

[2] This statement refers to the harvesting yards used in the specific case study as this does not always occur in the many different operational situations.

Table 4.6 Decision ranking
matrix adopted for criteria
weight combination

	A	B	C
A	A	A	B
B	A	B	C
C	B	C	C

Based on these criteria, it emerged that scenarios in which the cost of the production of trees to be chipped was lower were preferable for all cultivation models that produce biomass with a limited input requirement.

Similar considerations can be made for chip storage; scenarios in which this does not occur or where it occurs in a limited way have a lower environmental impact since they do not require the realization of dedicated structures and additional costs for moving the biomass. Finally, the use of transport machinery with a high volumetric load and low power supply from the engines was preferable due to the lower number of journeys involved and, consequently, the lower CO_2 emissions for transporting the same load.

Each of the thirteen criteria identified contributed equally to the overall assessment of the final scenarios selected (Y_n;H_n;L_n). The specific determination of the criterion was carried out using three classes identified by the letters A, B, C, with quality, quantity or capacity decreasing from A to C. The criteria were subsequently used to realize the decisional matrix shown in Table 4.6.

4.5 SRC and MRC Chains Assessment

The best solutions that can be implemented according to the MCA analysis process were defined by comparing all the hypothetical scenarios. The following method was used for this procedure:

1. Assessment of the comparison criteria through direct calculation and assignment to class A, B, C (see Table 4.7);
2. Assessment of all the Y_n; H_n; L_n scenarios using the product of the values obtained by the criteria set for both the environmental and economic aspects of each phase of the chain, calculated according to the decisional matrix in Table 4.6;
3. Realization of all the combinations between the scenarios proposed with the attribution of the overall value using a similar calculation procedure to that given in point 2.

The calculation of the decisional matrix involved combining the result obtained from the product of the two assessments with a third, without calculating the commutativity of the same. Given the high number of comparative variables for each scenario (cultivation, harvesting and logistics), it was deemed opportune to introduce an additional comparative procedure to give the analysis greater significance. This involved attributing three weights, respectively 3, 2, 1 to the A B C

Scenario	Environmental score		Economic score		Evaluation
Table 4.7 SRC and MRC scenarios evaluation throughout MCA					

Scenario	Environmental score		Economic score		Evaluation
Evaluation of harvesting scenarios					
H1	8. B	B	12. A	A	**A**
	9. B		13. B		
H2	8. A	B	12. C	B	**B**
	9. C		13. A		
H3	8. C	C	12. A	B	**C**
	9. B		13. C		
H4	8. B	C	12. C	C	**C**
	9. C		13. B		
Evaluation of crop management scenarios					
Y1	1.	B	A	10. B	**A**
	2.	B			
	3.	A			
	4.	A			
Y2	1.	C	C	10. C	**C**
	2.	C			
	3.	C			
	4.	C			
Y3	1.	A	A	10. A	**A**
	2.	B			
	3.	A			
	4.	B			
Y4	1.	B	B	10. B	**B**
	2.	B			
	3.	B			
	4.	B			
Evaluation of logistic scenarios					
L1	5.	A	A	11. B	**A**
	6.	C			
	7.	A			
L2	5.	B	A	11. C	**B**
	6.	A			
	7.	B			
L3	5.	C	B	11. A	**A**
	6.	B			
	7.	A			
L4	5.	C	C	11. C	**C**
	6.	A			
	7.	C			
L5	5.	A	A	11. C	**B**
	6.	B			
	7.	B			

assessments obtained from the criteria that contributed to the overall determination of the score of the individual scenarios Y_n; H_n; L_n. By substituting the assessments in "letters" with the corresponding numerical system, numerical coefficients were obtained through algebraic addition and averages that represented the environ-mental and economic aspects which, in turn, were added and averaged to obtain a single value that was representative of the hypothetical chain. The process described made it possible to assess in more detail all the elements contributing to the determination of the overall result for the chain.

By combining all the scenarios, 60 different chains were identified. A further selection was made by examining only those with an overall "A" assessment which resulted in a final sample group of 23 chains.

These, in turn, were classified using the numerical assessment method previ-ously described, and ordered according to decreasing value as shown in Table 4.8 .

By analysing the results obtained, it emerged that the best management–cultivation techniques included a marked frequency of the Y1L scenario (short rotation coppice, low input) with nine chains out of a total of 20, followed by the Y3L scenario (medium rotation coppice, low input).

The "preferred" harvesting yards were those included in the H3 and H1 sce-narios. However, it should be noted that not all the machinery can be universally used in every context, resulting in the removal of chains 23 (Y2-L1-H3) and 31 (Y2-L2-H3), and the possible combinations between harvesting yard H2 and cultivation scenarios Y3 and Y4. This was necessary for chains 23 and 31 since, to date, no cases have been found in the relevant literature of harvesting experiments using forestry machinery in high density SRC tree farms. This does not, however, mean that they cannot be effectively and suitably realized. The other possible combina-tions were excluded due to the technical impossibility of using cut-and-chip har-vester on stems with a butt diameter of more than 80–120 mm. In the fifth and sixth places in the classification were the chains that involved harvesting with an H2 yard. The overall analysis of these yards showed that they were not among the potentially best yards. This can be attributed to the decisional criteria imposed which took into account the investment necessary and the power supplied by the machinery which, in the specific case of the cut-and-chip harvester, was quite high, thus prejudicing the assessment. However, it should be pointed out that the use of these methods on a large scale would undoubtedly lighten the investment and harvest costs and, pre-sumably, the harvesting activities would also have a lower impact.

With regard to logistics, it is possible to note a substantial uniformity in terms of the frequency of repetition with an almost constant representativeness in all the scenarios except for L4. One of the best is L1 due to the transport distances of almost zero.

In order to identify the best possible chains it was necessary to carry out a further selection, therefore, using "numerical classification" as the discriminant, the study concentrated on the first five nos. 9, 45, 1, 41, 10.

This showed that the best scenarios involved:

- Management models with limited energy input requirements for both SRC and MRC tree farms;
- H1 type harvesting yards for SRC tree farms and H3 for MRC plantations;

Table 4.8 Classification of the best hypothesized chains by MCA analysis

Chain	Scenarios			Environmental score									Total	Economical score				Total	Total Score
				Y										Y	L	H			
9	Y1	L3	H1	B	B	A	A	C	B	A	B	B	2.22	B	A	A	B	2.50	2,36
45	Y3	L3	H3	A	B	A	B	C	B	A	B	B	2.11	A	A	A	C	2.50	2,31
1	Y1	L1	H1	B	B	A	A	A	A	A	B	B	2.33	B	B	A	B	2.25	2,29
41	Y3	L1	H3	A	B	B	B	A	B	A	B	C	2.22	A	B	A	C	2.25	2,24
10	Y1	L3	H2	B	B	A	A	C	B	A	C	A	2.22	B	A	C	A	2.25	2,24
2	Y1	L1	H2	B	B	A	A	C	A	A	A	C	2.33	B	B	C	A	2.00	2,17
5	Y1	L2	H1	B	B	A	B	A	B	B	B	B	2.33	B	C	A	B	2.00	2,17
17	Y1	L5	H1	B	B	A	A	B	A	A	B	B	2.33	B	C	A	B	2.00	2,17
43	Y3	L2	H3	A	B	B	B	A	B	A	A	B	2.22	A	C	A	C	2.00	2,11
49	Y3	L5	H3	A	B	B	B	A	B	B	C	B	2.22	A	C	A	C	2.00	2,11
55	Y4	L3	H3	B	B	A	B	B	C	A	C	B	1.89	B	A	A	C	2.25	2,07
13	Y1	L4	H1	B	B	A	A	B	A	A	B	B	2.11	B	C	A	B	2.00	2,06
6	Y1	L2	H2	B	B	A	A	B	B	B	A	C	2.33	B	C	C	A	1.75	2,04
18	Y1	L5	H2	B	B	A	A	B	B	B	A	C	2.33	B	C	C	A	1.75	2,04
47	Y3	L4	H3	A	B	A	B	B	C	A	C	B	2.00	A	C	A	C	2.00	2,00
51	Y4	L1	H3	B	B	B	B	B	A	B	C	B	2.00	B	B	A	C	2.00	2,00
29	Y2	L3	H1	C	C	C	C	C	B	C	B	B	1.56	C	A	A	B	2.25	1,90
53	Y4	L2	H3	B	B	B	A	B	B	B	C	B	2.00	B	C	A	C	1.75	1,88
59	Y4	L5	H3	B	B	A	B	B	B	B	C	B	2.00	B	C	A	C	1.75	1,88
21	Y2	L1	H1	C	C	C	C	C	A	A	B	B	1.67	C	B	A	B	2.00	1,83

- Chip transportation within the farm (distances of <1 km) using farm trucks or truck and trailers over maximum distances of 100 km.

The evaluations obtained through the MCA analysis were not globally representative of the best conditions but referred solely to the criteria set as identification factors, therefore the results referred only to a theoretical condition which was certainly indicative of SRC and MRC tree farms but not universally representative.

4.6 Application of LCA for the Wood–Energy Chains Identified

The next phase of the multicriteria analysis was to assess the five best chains 9, 45, 1, 41, 10 identified by the MCA analysis from the viewpoint of emissions in terms of CO_2 equivalent in g/kWh and the CER Wh/kWh of fossil energy required during the process.

Gemis 4.5 software was used to determine the results for each chain [17].

The objective was to compare the emissions of CO_2 equivalent generated by the chains identified and, for each of these, to determine the quantity originating from the crop management, the harvesting phase and the transportation of the chips. The CO_2eq parameter expresses the cumulated quantity of CO_2 and other greenhouse gases produced during the production processes. The non-renewable fossil energy requirements during the chain phases were calculated using the CER index. The cumulated energy requirement (CER) is an indicator for energy systems that makes it possible to quantify the primary energy requirement for products and services as part of a LCA. The cumulative energy requirement basically indicates the environmental pressure connected with the energy use.

The LCA inventory phase was developed by referring to a theoretical area of 10,000 m^2, listing the requirements during planting, cultivation, harvesting and transport, as stated in Table 4.9. The following data were taken into consideration for the LCA:

- Average annual production in kg_{db}/ha/year;
- Duration of tree farm in years;
- Fuel consumption expressed in kWh/ha/year for planting, cultivation, harvesting, transport;
- Water consumption in l/ha/year;
- Quantities of herbicides and pesticides in kg/ha/year;
- Quantities and types of fertilizers used in kg/ha/year;
- Type of means and distances of transport.

The inputs for each phase were calculated and estimated based on the information available from the relevant literature [2, 4–6, 9, 37, 38]. Fuel consumption during planting, cultivation and harvesting was given in "annual average values" expressed in kWh/ha/year, and was obtained by adding up the requirements of each cultivation operation or harvesting phase distributed over the years of the tree farm's life. This was necessary in order to identify the annual average value of

Table 4.9 Inventory data for the five most sustainable SRC and MRC chains according to the MCA

Processes	Units	9	45	10	1	41
Yield	kg$_{db}$/ha/year	12,000	14,000	12,000	12,000	14,000
Crop lifetime	years	10	15	10	10	15
Plantation						
Fuel consumption	kWh/ha/year	122.00	186.26	122.00	122.00	186.26
Type of fertilisers organic or mineral	–	Organic	Organic	Organic	Organic	Organic
Fertilisers (for each) herbicides	kg/ha/year	5,000	5,000	5,000	5,000	5,000
Water	l/ha/year	40	–	40	40	–
Pendimethalin + Linuron + Alachlor	kg/ha/year	0.08 + 0.04 + 0.14	–	0.08 + 0.04 + 0.14	0.08 + 0.04 + 0.14	–
Cultivation						
Fuel consumption	kWh/ha/year	417.00	270.36	417.00	417.00	270.36
Type of fertilisers	–	Organic	Mineral 8:24:24	Organic	Organic	Mineral 8:24:24
Fertilisers pesticides	kg/ha/year	8,000	160	8,000	8,000	160
Water	l/ha/year	600	360	600	600	360
Fenitrothion + Chlorpyrifos + Cypermethrin	kg/ha/year	0.84 + 0.36 + 0.06	0.50 + 0.22 + 0.04	0.84 + 0.36 + 0.06	0.84 + 0.36 + 0.06	0.50 + 0.22 + 0.04
Harvesting						
Fuel consumption	kWh/year	99.18	120.14	91.89	99.18	120.14
Transports						
Type of mean		Truck trailer	Truck trailer	Truck trailer	Rural trailer	Rural trailer
Distance	km	100	100	100	1	1

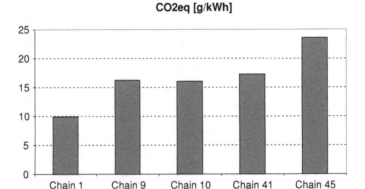

Fig. 4.2 CO_2 $_{eq}$ emissions for the most suitable SRC and MRC analysed chains

emissions and energy consumption since the type of cultivation intervention differed over the years. Similar considerations were made for all the other operations in the cultivation calendar, specifying for each one the quantity and type of material used as described in Table 4.9 and 4.2 in Sect. 4.3.1 which also gives the working capacities of the machinery.

With regard to fuel consumption for machinery, reference was made to a specific consumption of diesel of between 0.225 and 0.250 kg/kWh which varied in relation to the machinery which had a power utilization coefficient that also varied between 40 and 80%. With regard to the characteristics of the diesel used in the farm machinery and for the modelling of the logistical phase and energy utilization, the same coefficients given in Chap. 2 , Sect. 2.7, were used.

From an analysis of the total CO_2eq emissions, chain no. 45 (Y3-L3-H3) had the greatest impact, with a total of 23.567 g/kWh, whilst chain no. 1 (Y1-L1-H1) was the most sustainable with 9.874 g/kWh. The other three chains identified, nos. 9, 10, 41, presented similar values of around 16.500 g/kWh (see Table 4.10 and Fig. 4.2 outlines the emissions of the five combinations selected).

In all the chains identified, field operations emerged as having the greatest impact, followed by the logistical and harvesting phases. In percentage terms, the three phases showed a variable incidence of cultivation techniques of between 46 and 85%, from 0.5 to 39% for logistics and, finally, from 11 to 24% for harvesting.

With regard to harvesting yards, increased levels of mechanization resulted in a decrease in the percentage incidence in terms of CO_2eq with respect to the other phases of the cultivation scenario.

The results of the CER index shown in Fig. 4.3 and Table 4.11 confirmed a similar trend in the CO_2eq analysis for chains no. 45 and 1, whilst the trends changed for chains no 9, 10, which displayed primary energy requirements close to those of chain no. 45 due to the higher demand for fossil fuels caused by the greater transport distances.

Table 4.10 CO_2 eq emissions for the five most suitable SRC and MRC chains identified trough the MCA

Scenario		9 Chain		45 Chain		10 Chain		1 Chain		41 Chain	
Total	$gCO_2/$ kWh	**16.224**		**23.567**		**16.052**		**9.874**		**17.217**	
Y											
Crop Management	%	Y1	46.0	Y3	62.1	Y1	46.5	Y1	75.6	Y3	85.0
L											
Logistic	%	L3	38.9	L3	26.8	L3	39.3	L1	0.5	L1	0.3
H											
Harvesting	%	H1	15.1	H3	11.1	H2	14.1	H1	23.9	H3	14.8

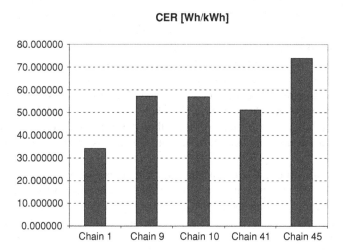

Fig. 4.3 CER requirement for the most suitable SRC and MRC analysed

Table 4.11 Result of CER index for the five most suitable SRC and MRC chains

	9 Chain	45 Chain	10 Chain	1 Chain	41 Chain
Wh/kWh	57.292	74.023	56.944	34.269	50.999

4.7 Wood Biomass from Forestry

Due to high availability and versatility of use, wood biomass from forestry is enjoying steady growth in the complex field of renewable energy.

The continuing increases in the price of oil over recent years have stimulated the diversification of traditional products obtainable from the forest; forestry firms are devoting more attention to streamlining production techniques and to the commercial exploitation of the obtainable assortments. However, the number of forestry firms that are organizing to set up yards capable of harvesting what up to now has been considered "production scrap" is still limited.

In this second example, we use MCA and LCA to study some wood and chip production chains for heat generation in heating systems.

4.8 Multicriteria Analysis for Definition of Forestry Scenarios

As opposed to short and medium rotation coppices, the production of wood biomass in the forest involves more complex production processes due essentially to the difficult orographic characteristics of the working environment. These conditions hinder the generalized introduction of mechanical equipment in the forestry chains, therefore it is not always possible to use the machinery that would contribute to simplifying the processes and spreading the use of "clean energy" in an economic way. Another problem is the lack of specific technologies for all silviculture typical of the Mediterranean basin. On the one hand there are huge, advanced technology machines designed and developed for working in large Northern European and American forests where there are fewer difficulties related to the morphology of the land; on the other there are still few small-medium solutions that would simplify and facilitate operations in the more difficult to access forests.

Further considerations concern the type of biomass that can be produced; the new focus on the differentiation of goods and services offered by forestry firms raises necessary questions about what to produce. In Italy much of forest biomass production is still oriented towards firewood because it fetches better prices on the Italian market, prices that range between 80 and 130 €/t.

Independently of the technologies used and the type of biomass that can be obtained, the production phases consist of: cutting the plant, processing into commercial assortments, yarding-extraction and transportation to the final user.

All of these phases can be managed with many yards and methods, but they are all restricted by a time span. Felling and logging must be completed before vegetation resumes and that means, basically from fall to spring. Therefore, it is essential to identify the technologies and working models that make it possible to produce the greatest quantity of biomass at the lowest cost during the working season.

On the basis of the considerations set out for the definition of possible chains, we took into account and then combined different solutions for each scenario. As the analysis described in Sect. 2.1 explains, the study is based on a comparison between a series of criteria related to environmental and economic aspects for the chains that could be used in the production of forest biomass. The criteria concern the harvesting phase, transportation, storage and the use of wood and chips to produce energy (heat).

The possible scenarios that can be set up for the production of wood biomass were developed with reference to biomass obtained from class 1 managed coppices I[3] [19] in average phytosanitary conditions comprising mainly oak, alder, ash and black hornbeam, hypothesizing ordinary cutting with a 20-year rotation and release of 100 mother plants per hectare from which it would be possible to obtain approximately 90 t/ha of firewood, 140 t/ha of chips or 100 t/ha of wood and 30 t/ha of chips when branches are collected. The given amounts of available biomass should be considered as mean values obtained in trials conducted in the Apennines in north-central Italy [13, 32, 39].

4.8.1 Description of Harvesting Yard

The technologies that can be used to produce wood biomass depend on multiple variables including the form of forest management (coppice, coppice with standards, high forest) which determines the treatments, that is the operational methods followed for cutting or felling, the accessibility and morphology of the land that determine the systems that can be implemented and the obtainable products.

In addition to identifying the technologies that can be employed, the harvesting system requires a definition of the processing systems. In Italy, even today, forestry firms work with relatively low levels of mechanization and the short wood system is still widely used.

This method is used for producing firewood for stoves for home heating. It calls for processing on the stump and the extraction of the wood in commercial assortments. The potential advantages include the possibility of using light extraction equipment, and the possibility of adapting the low investments for acquiring technologies for difficult or inconvenient landings. The disadvantages are reduced daily productivity which means concentrating on the main products and incomplete exploitation–utilization of the biomass that is left in the forest.

Contemporaneous production of several commercial assortments (wooden poles, firewood, chips) can be readily achieved with the full tree system that calls for felling and extracting the tree together with the branches and then it is processed and/or chipped at the landing. This method is used when full trees are used for chip production, when the ground has to be cleared of branches to reduce risks of fire or, in general when there are plans to collect branches.

The advantages consist of greater operational ease and safety and the simple utilization of all the available biomass; on the other hand, this method requires more spacious landings and logging equipment that is more powerful than standard farm tractors to move big loads. The full tree system also means having to concentrate several activities at the landing, thereby requiring good organization of the

---

[3] Forests with gradients ranging from 0 to 20% where extraction is possible in all directions using commonly available technologies.

Table 4.12 Main features of the hypothesized forestry harvesting yards

Scenario	Yard mechanisation	Biomass produced	Biomass production cost (€/t)
Y 1	*Felling-processing* chainsaw	Firewood	55
	Extraction farm tractor, equipped with front and rear container bins		
Y 2	*Felling* chainsaw	Firewood and Wood Chips	90
	Extraction whole trees, agricultural tractor with winch		
	Processing at the landing with chainsaw		
	Cipping mobile chipper 150 kW with crane		
Y 3	*Felling* chainsaw	Wood Chips	47
	Extraction whole trees, at the landing with cable crane		
	Cipping mobile chipper 150 kW with crane		
Y 4	*Felling* chainsaw	Firewood & Wood Chips	60
	Extraction whole trees, at the landing with cable crane		
	Processing at the landing with excavator-based harvester		
	Cipping mobile chipper 350 kW based on truck trailer with crane		

equipment and of the logging and processing crews to prevent accidents, interference and delays [13, 32, 39].

On the basis of the considerations presented for managing harvests in coppice forests we have defined the harvesting yards shown in Table 4.12 and designated Y1, Y2, Y3, Y4. They are differentiated by the type of equipment used, the type of biomass that can be obtained (firewood, chips or both), investment costs, working capacity and versatility of use.

For each yard, we determined: hourly output, total Kilowatt power of the equipment, cost per ton of product and the necessary investment. The cost per ton was determined analytically on the basis of the references available in the literature [13, 14, 16, 19, 29, 32, 39]. The investment amounts were based on the price lists available on the web and from prices published in dedicated scientific journals [20–22].

The creation of chains for collecting biomass requires careful planning of the entire production process. And, since it is considered a secondary business activity it is essential that it not have a negative impact on the productivity and profitability of the main business.

Currently, residual biomass can be managed with two types of collection logistics:

- *Mobile yards:* the machinery is moved to the biomass
- *Fixed yards:* residual branches are moved to the chopping–shredding unit.

A further difference concerns the methods of biomass collection that can be done by bundling or with chopper–loaders. Bundling makes it possible to compact scrap/waste (branches and tops) to obtain, evenly shaped, high density bales to optimize transport and handling, as well as storage thanks to the reduction of deterioration causing phenomena. Bundling is a widespread and efficient process in farming where the orographic conditions of the plots permit easy collection. In forests, limited accessibility makes collection with these methods difficult and often uneconomical. Therefore, the material is often abandoned on the ground and becomes a potential source of risk of both fires and the spread of diseases. The possible solutions for the forest include industrial bundling with dedicated forest equipment. These machines, that are set up on forwarders working in the forest or on trucks at the landings, guarantee production of approximately 10 bales/h corresponding to 2–7 t/h of wood [31].

Currently, due to the lack of specific technologies and the technical inadequacies of the machinery in the forest continuous chopping in the forest is not common in Italy. However, in Northern European forests where industrial silviculture makes it possible, chips can be collected thanks to forwarders equipped with felling heads, chopping devices and temporary product storage systems (Valmet 801 Combi BioEnergy, 2010).

In many European countries the most common solution for collection is the use of chippers or chopping machines mounted on trucks. This method, that can be readily implemented when working with the long wood system, requires suitably sized landings located near the main road so as to facilitate the operations and simplify moving the equipment and hence the biomass [32].

The forest yards hypothesized for the production of wood biomass are described below. Yard Y1 represents the first level of mechanization that can be implemented and it is the most common management model for harvesting firewood in Italy. Cutting–processing is semi-mechanical, done by two operators who cut the wood into 1 m pieces using medium-powered 3–5 kW chain saws in the forests on stumps. Extraction is done with 75 kW farm tractors modified to work in the forest, equipped with front and back metal cribs, with a mean load capacity of 3.5–5 mst of wood. The mean daily output is determined from the slowest phase which is extraction with values ranging from 2.5 to 4.5 t/day (7 h working hours per day) [13, 16, 32, 39].

The other yards use the full tree system (FTS). In yard Y2 cutting–processing is done by four operators using chain saws. One 75 kW tractor equipped with a two-drum winch, each with 80 m of cable with a tractive force of 50 kN is used to extract full trees at the landing. The forest winch is a simple machine consisting of four basic parts: a drum on which the cable is wound; a mechanical or hydraulic transmission, a clutch and support frame. Using these machines makes it possible to gather the logs in groups even from areas that are difficult to reach with mechanical equipment thereby increasing both productivity and safety at the same time [28, 30].

Once the processing into firewood is completed at the landing, the next phase is chopping the residue with a medium power 150 kW chipper mounted on a two-

Fig. 4.4 *Left*: Big power chipper mounted on a four axles truck. *Right*: Medium power chipper mounted on a two-axle trailer

axle trailer equipped with a crane that can handle trees with a base diameter of 0.4 m, with a mean hourly output of 8–12 t_{wb} Fig. 4.4 [13, 16, 32, 39].

Chipping material that has been skidded involves greater risks of contamination with the ground, bringing in impurities that can damage the chipping equipment. Therefore, it is preferable to opt for solutions that masticate rather than cut the material in order to avoid continuous maintenance/repair work that would impact both overall operating costs and product quality. The masticating machines process the biomass with rotating rollers equipped with dull devices as opposed to the chippers that cut the wood with several blades.

The use of chippers to process broadleaved trees, characterized by extremely heterogeneous growth patterns, with many large diameter primary and secondary branches necessitates preliminary cutting and therefore, higher costs and emissions.

Yards Y3 and Y4 differ as to extraction methods, and use a powered stationary cable crane equipped with an autonomous 80 kW motor, 15 mm carrying diameter and a length of 500 m. The cable cranes are machines that can be used to advantage when the slopes exceed 30%, when there is little primary and secondary infrastructure or when the orgraphy and forest ground make it impossible to bring in mechanical equipment. The market offers a wide range of powered stationary cable cranes, and mobile cable cranes are the most widely used at this time. These cranes have a variable number of cables wound on drums that are driven by autonomous motors or using the power from a tractor, and they are quickly set up and taken down. They can be used for uphill or downhill extraction.

The biomass is harvested, collected and extracted using a carriage connected to the cable: during the work phase the carriage is locked on the carrying line at a transfer station where one or more operators hitch the material.

In addition to the advantages described, the use of a cable crane for extraction makes it possible to minimize damage to the ground and to increase productivity in spite of the relatively long times needed for set up and the medium–high investments that demand sufficient quantities of wood to cover operating, set up and dismantling costs [28, 31].

The procedure at yard Y3 calls for cutting the trees by two operators with chainsaws adopting measures to direct the fall of the trees towards the area covered by the cable crane to facilitate hitching and extraction. Once the unloading station, situated uphill from the cutting area has been reached, the full trees are first cut by an operator with a chainsaw and then reduced into chips by a chipper similar to the one used in yard Y2.

Yard Y4 is the most highly mechanized for the four hypothesized: felling is done by two chainsaw operators, extraction is done with a light, powered stationary cable crane. Processing is done with 70 kW 12 t excavator, equipped with a felling–processing head. The full trees brought to the unhitching site by the crane are unloaded, and grasped by the forest processor that cuts them into 1 m sections; the scrap is cut up by a 350 kW chipper mounted on a truck able to process wood with a maximum diameter of 0.65 m at a mean hourly rate of 10–18 t_{wb} Fig. 4.4 [13, 16, 32, 39].

For the harvesting phase it seems essential to optimize the least productive phase which is usually extraction because the use of low capacity equipment can make the use of high-productivity machinery for the other processes—and hence the entire process of biomass production—inefficient.

To conclude, the extreme heterogeneity of the forests makes it possible to adopt multiple solutions for producing biomass. Therefore, in each situation it is essential to assess the critical points and the available opportunities to create well-organized chains that exploit the available technologies to the fullest to guarantee good productivity, lower costs and as a result, the greater use of low-impact biofuels.

4.8.2 Description of Logistic Scenarios

As for the definition of the harvest scenarios, logistical choices must be made on the basis of a case-by-case assessment of the contingent situation. There are many variables that contribute to the identification of the most efficient logistics in which the transported load, hourly costs of the vehicle, type of road, etc. play a primary role .

The transportable load depends on both the technical features of the vehicle (payload, maximum volume capacity), and road regulations which impose maximum load and speed limits which, in turn, affect the overall operating times.

The many studies conducted on this subject have made it possible to define the following points [32]:

- It is economically more advantageous to transport stem wood as opposed to chips or bundles;
- It is economically more advantageous to transport chips as opposed to tree sections;

Table 4.13 Main features of the hypothesized logistics scenarios for the forestry wood energy chains

Scenario	First transport	First storage	Second transport	Second storage	Type of vehicles
L1	0	Dedicated areas	15	–	Rural trailer
L2	5	Dedicated areas	80	Plant storage	Rural trailer + truck trailer
L3	15	Dedicated areas	15	–	Rural trailer + rural trailer
L4	10	Dedicated areas	40	Plant storage	Rural trailer + truck trailer
L5	100	–	–	Dedicatea areas	Truck trailer
L6	20	Dedicated areas	20	Plant storage	Truck trailer + truck trailer

- It is economically more advantageous to transport tree sections than tops and branches;
- It is always advantageous to opt for the faster vehicle with greater capacity compatibly with road and landing site conditions.

Other aspects to consider concern the characteristics of the road and landings. Good secondary and primary roads and landings guarantee easier movement of equipment which is essential for optimizing the logistic chains.

Since transport is done by mainly tired vehicles, the development of wood–energy chains requires careful planning and considerations regarding the impact on local traffic. This is important in the development of local chains characterized by roads that are either poor or cannot technically withstand high load capacity vehicles [34].

The hypothesized logistic scenarios call for the use of two types of vehicles: tractors with farm wagons or truck and trailers. In Italy, farm and forestry firms generally use farm trailers to transport wood and biomass since they are low cost vehicles found in all farming and forestry contexts. However, it is important to note that the low load capacity and low speeds of farm trailers can make the entire production process uneconomical.

The use of high payload vehicles such as the truck and trailer increases the economic advantages for long distance transport, but on the other hand, these vehicles require suitable roads near the harvesting site for prompt transfers. Another option is the use of containers. This system offers good load capacity, high operational versatility and reduces the need to depend on the equipment at the yard thereby reducing dead times; on the other hand it requires greater investments in the trucks and containers [27].

Six solutions have been defined for the logistic scenarios, identified as L1, L2, L3, L4, L5, L6 that are characterized by the distance covered, number of storage operators for processed material, load capacity and power of the vehicles used Table 4.13.

As in the case study on short rotation coppices, the scenarios were developed to show solutions typical of simple situations in which the biomass is used in small, home heating systems and in complex cases that involve the use of medium–large boilers to heat more than one structure or building at the same time.

Scenarios L2, L4, L6 call for the delivery of the biomass to a conversion plant located between 40 and 100 km from the collection site. These three logistic models presume that the biomass is first stored in dedicated areas near the collection area situated within a 20 km radius of the harvest zone and then transported to the final destination.

Scenario L1 is the simplest logistic model; once collected in the dedicated area, the biomass is used in a heating system situated not more than 15 km away. Scenario L3 hypothesizes an intermediate situation, with a single storage of the biomass situated at 15 km between the production area and delivery to the final user at a distance of 15 km. Finally, in scenario L5, the biomass is transported over a distance of 100 km to dedicated areas from which it will be moved to a conversion system.

The storage phase is often not carefully evaluated and frequently presents yet another issue to resolve. The problems are due mainly to the quali-quantitative deterioration of the chips with an estimated monthly loss of 3–4% and to the reduced availability of storage space [34]. These aspects have an economic impact on the production processes, therefore, they were assessed by considering the number of times the biomass is stored.

4.8.3 Description of Energy Plants

For the definition of the energy utilization scenarios we considered five different types of heat-producing systems identified as E1–E5 with powers ranging from 20–500 kW Table 4.14. Only scenario E1 calls for the use of firewood, the other systems are fueled with chips. The comparison parameters take into account the efficiency of the conversion, the technological and management/operational complexity of the system and finally the ratio between investment cost and kW produced by the boiler [35].

Some important data and considerations that are also applicable to this case study on systems fueled with biomass are presented in Sect. 3.3.3 Chap. 3. One important aspect to consider for efficient energy conversion is the quality of the chips, which is determined by quantifying the fibre and water content and chip size. Fibre content indicates the percentage of wood from which, using special calculations that take increase and decrease factors due to bark and leaves into account, it is possible to calculate the lowest thermal value of the PCI wood.[4]

---

[4] Thermal value is the amount of energy that is potentially available from the complete combustion of one unit of weight (kg) of wood. This parameter considers water released as vapour and hence net of the heat energy needed for the evaporation of the water contained in the wood (0.68 kWh/kg H_2O).

Table 4.14 Energetic plants features

Scenario	Plant typology	Fireside efficiency (%)	Investment costs (K€)	Biomass input	Technological and management complexity
E1	i.e. 20–60 kW direct combustion private plants using firewood for heat production	70	7–12	Firewood	Low
E2	i.e. 20–60 kW direct combustion private plants using wood chips for heat production	80	15–30	Wood chips	Low
E3	i.e. 100–150 kW direct combustion private plants using wood chips for heat production	89	80–120	Wood chips	Medium
E4	i.e. 300–350 kW direct combustion private plants using wood chips for heat production	85	180–220	Wood chips	Medium
E5	i.e. 500–550 kW direct combustion private plants using wood chips for heat production	90	300–350	Wood chips	High

The water content is the percentage of water in the wood; good chips should have approximately 30% water to guarantee efficient energy conversion [16]. Water has a negative impact on combustion: the greater the amount of water in the wood the higher the energy costs due to evaporation. And, there are greater possibilities of difficult lighting/-starting, incomplete combustion, sudden extinguishing, and non optimum performance which lead to higher costs for acquiring the technologies to operate more complex combustion processes.

The size of the chips is also important: the dimensions must meet the system's specifications in order to prevent blockage of the automatic charging system and irregularities during the combustion phase that would reduce the overall performance of the system.

For these scenarios we referred to heating systems that are operating in Tuscany [2, 39] and other regions in Italy that are typical of simple solutions in which wood and chips are used for heating single homes, and other more complex situations in which the biomass is used to obtain heat energy in remote heating systems.

4.9 Wood Biomass Decision Criteria Identification

As explained in Sect. 4.2 for each scenario hypothesized for the harvesting, logistics and energy utilization scenarios we evaluated the decision-making criteria related to environmental impact and economic sustainability.

Table 4.15 Definition of evaluation criteria for forestry wood energy chains

Decision criteria	Weights		
	A	B	C
Environmental Aspects			
kW Harvesting yard	≤200	>200 and ≤350	>350
No. of machine	≤3	>3 and ≤5	>5
Distance (km)	≤40	>40 and ≤50	>50
Type of mean	Truck trailer	Truck trailer + rural trailer (mixed)	Rural trailer
kW Transport	150	>150 and ≤250	>250
Fireside efficiency	>80	>70 and ≤80	≤70
Economic aspects			
Production cost (€/t)	≤50	>50 and ≤75	>0,75
No. of storages	–	1	2
Investment (k€/kW)	≤0.50	>0.50 and ≤0.65	>0.65
Technological and management complexity	Low	Medium	High

Specifically, we defined seven environmental and four economic criteria. The definitions were based on the following considerations:

- The preferable scenarios are those in which fewer machines are used in harvesting and the vehicles/equipment in the harvest yard use the lowest total power.
- Environmental impact is lower if the biomass is transported with low-power vehicles, if the transport is over short distances with high-load capacity vehicles and if the conversion systems have high transformation efficiency.
- Economic benefits are possible when the differences between the costs of producing and managing the biomass, between the investment costs and power of the systems are minimal, when storage operations are reduced and when management and maintenance of the systems are simple [35].

As to the choices for the decision-making criteria related to the environmental aspects, the considerations in Chap. 2 Sect. 2.4 can be taken as valid.

In addition to the pressure indicators related to collecting agricultural residues, there are many other indicators that play a priority role in the evaluation of environmental impact for the forestry industry, such as rural development (e.g. quality local jobs); economy of the region (e.g. timber processing capacity); biodiversity conservation; water management; landscape; informal local recreation; economic regeneration of urban areas; tourism; green networks in settlements (e.g. linking town and country); habitat networks; climate change; biomass for energy and community involvement (e.g. local community groups). The need to summarize the multicriteria approach in this chapter limited the analysis to the study of the sustainability of the forest biomass production processes through the identification of the pressure indicators shown in Table 4.15, which are considered most representative of the forest wood biomass production chains.

Table 4.16 Results of the MCA analysis for the harvesting yards, logistic and energetic scenarios

Scenario		Environmental score		Economics score	Evaluation
Harvesting					
Y1	1.	A	A	8. B	A
	2.	A			
Y2	1.	B	C	8. C	C
	2.	C			
Y3	1.	B	B	**8. A**	A
	2.	B			
Y4	1.	C	C	8. B	B
	2.	B			
Logistic					
L1	3.	A	A	9. B	A
	4.	C			
	5.	A			
L2	3.	C	C	9. C	C
	4.	B			
	5.	C			
L3	3.	A	B	9. B	B
	4.	C			
	5.	B			
L4	3.	B	C	9. C	C
	4.	B			
	5.	C			
L5	3.	C	B	9. B	B
	4.	A			
	5.	B			
L6	3.	A	B	9. C	C
	4.	A			
	5.	C			

In addition to meeting the requisites for environmental sustainability, planning the chains for the supply and transformation of biomass into fuel requires good organization and efficient use of the available technologies. This is essential because decisions that are not carefully made can nullify all the advantages which could be obtained from the generalized use of the technologies and consequently the advantage of using alternative fuels.

4.10 Forestry Wood–Energy Chains Assessment

The best solutions to implement are determined by applying the procedure presented in Sect. 4.5. The evaluation was done considering the environmental and economic criteria as identical in terms of significance.

We obtained the assessments shown in Tables 4.16 and 4.17 by applying the environmental and economic criteria for the harvesting, logistic and energy transformation scenarios .

Table 4.17 Results obtained applying the MCA to the energy plants hypothesised

Energetic scenario	Environmental score	Economics score			Evaluation
E1	7. C	10.	A	A	B
		11.	A		
E2	7. B	10.	B	A	A
		11.	A		
E3	7. A	10.	C	C	B
		11.	B		
E4	7. A	10.	B	B	A
		11.	B		
E5	7. A	10.	B	C	B
		11.	C		

For the harvesting scenarios, the results show, as in the collection of residues from pruning in olive groves, an opposing curve between economic and environmental sustainability. The yards with lower levels of mechanization are preferable according to environmental criteria, but they are often unsustainable in terms of costs and daily productivity.

The yards where two commercial assortments are products (firewood and chips) are economically and environmentally worse than the others because of the need for more processes that require more machinery and hence involve higher costs. In any case, if appropriately planned, the collection of processing scrap can contribute to complementing the profits obtainable from the sale of the main products.

At the sites where the trees are completely chips there is high productivity and better economic results, but from the environmental standpoint the assessments are not excellent because they require high capacity machinery which inevitably involves greater consumption of fossil fuels.

The same statements relative to the logistics in Sect. 3.3.2 of Chap. 3 and Sect. 4.3.3 of this chapter apply to the logistics of transporting forest biomass. The use of vehicles with low load and volume capacity over long distances increase the economic and environmental of these processes.

The assessments concerning energy production from residue from olive tree pruning, i.e. the use of high output systems means more efficient energy use that translates into lower consumption of fossil fuels. In particular, there are good economic results in systems with less than 60 kW heating power, but the environmental side is not so good because there are minimum possibilities for regulating these systems which also present more functional instability, whereas in systems with 100-500 kW power, energy output is higher thanks to the higher thermal inertia and technical solutions that permit accurate control of the combustion parameters, but at the same time involve higher investments.

We identified 25 from the combination of the harvesting, logistic and energy conversion chains. They show that, according to the MCA analysis, scenarios Y3 and Y1 are preferable for harvesting, scenarios E2 and E4 for energy conversion

Table 4.18 Classification of the best forestry wood energy chains hypothesized trough MCA analysis

Chain	Scenarios			Environmental			Total	Economic			Total	Total score
				H	L	E		H	L	E		
1	Y1	L1	E1	AA	ACA	C	2,33	B	B	AA	2,5	**2,42**
3	Y1	L3	E1	AA	ACB	C	2,17	B	B	AA	2,5	**2,33**
5	Y1	L5	E1	AA	CAB	C	2,17	B	B	AA	2,33	**2,33**
36	Y3	L1	E3	BB	ACA	A	2,33	A	B	CB	2	**2,17**
8	Y2	L1	E2	BC	ACA	B	2,00	C	B	BA	2	**2,00**
10	Y2	L1	E4	BC	ACA	A	2,17	C	B	BB	1,75	**1,96**
37	Y3	L1	E4	BB	ACA	A	2,33	A	B	BB	1,5	**1,92**
35	Y3	L1	E2	BB	ACA	B	2,17	A	B	BA	1,67	**1,92**
59	Y4	L1	E2	CB	ACA	B	2,00	B	BB	BA	1,83	**1,92**
38	Y3	L1	E5	BB	ACA	A	2,33	A	B	BC	1,33	**1,83**
43	Y3	L3	E2	BB	ACB	B	2,00	A	B	BA	1,67	**1,83**
51	Y3	L5	E2	BB	CAB	B	2,00	A	B	BA	1,67	**1,83**
57	Y3	L6	E4	BB	AAC	A	2,33	A	C	BB	1,33	**1,83**
45	Y3	L3	E4	BB	ACB	A	2,17	A	B	BB	1,5	**1,83**
53	Y3	L5	E4	BB	CAB	A	2,17	A	B	BB	1,5	**1,83**
55	Y3	L6	E2	BB	AAC	B	2,17	A	C	BA	1,5	**1,83**
44	Y3	L3	E3	BB	ACB	A	2,17	A	B	CB	1,33	**1,75**
46	Y3	L3	E5	BB	ACB	A	2,17	A	B	BC	1,33	**1,75**
52	Y3	L5	E3	BB	CAB	A	2,17	A	B	CB	1,33	**1,75**
54	Y3	L5	E5	BB	CAB	A	2,17	A	B	BC	1,33	**1,75**
61	Y4	L1	E4	CB	ACA	A	2,17	B	B	BB	1,33	**1,75**
47	Y3	L4	E2	BB	BBC	B	1,83	A	C	BA	1,5	**1,67**
49	Y3	L4	E4	BB	BBC	A	2,00	A	C	BB	1,33	**1,67**
39	Y3	L2	E2	BB	CBC	B	1,67	A	C	BA	1,5	**1,58**
41	Y3	L2	E4	BB	CBC	A	1,83	A	C	BB	1,33	**1,58**

and finally, for logistics, model L1 is prevalent, followed by scenarios L3 and L5. Table 4.18 shows that the nine best combinations will be analysed according to the LCA.

4.11 Application of LCA for the Forestry Wood–Energy Chains Identified

For forestry chains, the analysis of CO_2 equivalent emissions and the CER was carried out using the same tools and coefficients as for the chains studied previously, making reference to the direct emissions emitted by the machinery during operation. Table 4.19 gives the main variables analysed for the inventory phase.

With regard to the harvesting yards, the results show that the increase in the level of mechanization corresponded to a substantial reduction in CO_2 equivalent emissions per ton of biomass produced Table 4.20. In particular, the yards with the

Table 4.19 Inventory data for the most sustainable chains identified through MCA analysis

Scenario	Unit	Chain								
		1	3	5	59	35	37	36	8	10
Harvesting		*Y1*	*Y1*	*Y1*	*Y4*	*Y3*	*Y3*	*Y3*	*Y2*	*Y2*
Yield	kg/year	4,500	4,500	4,500	7,000	7,000	7,000	7,000	7,000	7,000
Harvesting yard capacity	t/h	2	2	2	3.5	3.5	3.5	3.5	2.5	2.5
Operation time	h/year	3.75	3.75	3.75	4.75	4.50	4.5	4.5	6	6
Fuel consumption	kW/year	183.19	183.19	183.19	419	301.49	301.49	301.49	390	390
CO_2eq emissions	g/kW	115.69	116.370	121.370	51.835	47.923	45.162	43.599	53.440	49.339
Logistic		*L1*	*L3*	*L5*	*L1*	*L1*	*L1*	*L1*	*L1*	*L1*
Means typology	–	Rural trailer	Two rural trailer	Highway truck	Rural trailer	Rural trailer	Rural trailer	Rural trailer	Rural trailer	Rural trailer
Transport distance	km	15	15 + 15	100	15	15	15	15	15	15
Energetic		*E1*	*E1*	*E1*	*E2*	*E2*	*E4*	*E3*	*E2*	*E4*
Operation time	h/year	2,200	2,200	2,200	2,800	2,800	2,800	2,200	2,800	2,800
Life time	year	10	10	10	12	12	15	15	12	15
Power	kW	20–60 (40)[a]	20–60 (40)[a]	20–60 (40)[a]	20–60 (40)[a]	20–60 (40)[a]	300–350 (325)[a]	80–120 (100)[a]	20–60 (40)[a]	300–350 (325)[a]
Fireside efficiency[b]	%	70	70	70	85	85	85	89	85	85
Electricity from grid	kWh/kW	0.02	0.02	0.02	0.02	0.02	0.02	0.02	0.02	0.02
Plant weight (steel weight)	kg	500	500	500	1,500	1,500	3,000	2,000	1,500	3,000
Emission control technology	–	–	–	–	–	–	Cyclone	Cyclone	–	Cyclone
Ash produced	kg/kW	2.50%	2.50%	2.50%	2%	2%	4%	4%	2%	4%

[a] Average value taken as reference in the LCA assessment

[b] Efficiency measured at fireside. For the LCA, reference was made following these factors (chains 1,3,5 efficiency 70%; chains 59,35,8 efficiency 75%; chain 36, 80%; chains 37,10 efficiency 85%) listed in the GEMIS database software that takes into account the overall efficiency of the plants

Table 4.20 CO_2eq emissions for the nine most suitable forestry wood chains

Scenario		1 Chain	3 Chain	5 Chain	8 Chain	10 Chain	35 Chain	36 Chain	37 Chain	59 Chain
Total	$gCO_2/$ kWh	115.690	116.370	121.370	51.835	47.923	46.934	45.162	43.599	53.440
Y Cutting Harvesting Chipping	$gCO_2/$ kWh	52%	51%	50%	31%	34%	27%	28%	29%	32%
L Logistics	$gCO_2/$ kWh	1%	1%	5%	2%	2%	2%	2%	2%	2%
E Thermal energy production	$gCO_2/$ kWh	48%	47%	45%	67%	64%	72%	70%	69%	66%

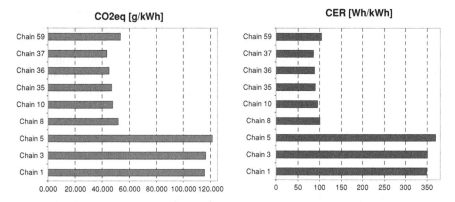

Fig. 4.5 CO$_2$ eq emissions and CER requirement for the most suitable forestry wood–energy chains analysed

Table 4.21 CER for the most suitable scenarios identified

Chain	1 Chain	3 Chain	5 Chain	8 Chain	10 Chain	35 Chain	36 Chain	37 Chain	59 Chain
W/kWh	348.6600	351.1300	369.2700	100.8200	95.1320	90.5630	88.1750	86.0680	104.2000

least impact were those that adopted the Y3 working model, i.e. where the trees obtained with the full tree system were integrally chipped. This working model certainly constitutes an interesting solution in terms of savings and productivity, but it is not easy to apply due to the frequently difficult conditions of accessibility in wooded areas.

The current working models, such as Y1, which are widely used in Italy and a large part of the Mediterranean basin, are unsustainable both economically and in terms of CO$_2$ equivalent emissions, with values that are double those recorded for yards with a high level of mechanization Table 4.20 and Fig. 4.5.

Logistics emerged as being the phase with the lowest impact compared to other phases; even in scenario 5, where the biomass was transported to distances 100 km away, the incidence did not exceed 5%.

The energy conversion phase for all the chains identified was inevitably responsible for the greatest CO$_2$ emissions, with values varying between 45 and 72%. According to the table, the E1 scenario—involving the use of firewood in domestic stoves—had the least emissions, but when compared to the overall value of CO$_2$, it became the most important element, not just in the hypothetical chains where it was present, but also with respect to the other energy utilization scenarios.

Analysing the more sustainable chains, no.37 was the best in terms of CO$_2$ equivalent as well as the CER Table 4.21 and Fig. 4.5, although it should be noted that it did not correspond to the preferable scenario obtained from the MCA analysis represented by no. 1. This was due to the multiple assessment criteria set by the two analyses which, despite having a common objective, i.e. the identification of the best chains, demanded different modalities and decisional variables.

It follows that the three best chains identified by the MCA analyses 1, 3, 5 were those with the highest impact due to the use of low energy efficiency boilers and the limited use of mechanization during the biomass production phase, which led to lower daily productivity, longer working times and, consequently, higher fuel consumption.

References

1. Alpenergywood, AA.VV. (2009) La produzione di biomasse legnose a scopo energetico. Technical report, Veneto Agricoltura
2. ARSIA. AA.VV. (2004) Le colture dedicate ad uso energetico, Quaderno ARSIA n.6
3. Assoverde. Prezzi informativi dei principali lavori di manutenzione e costruzione del verde e delle forniture di piante ornamentali, 2008/09
4. Balsari P, Manzone M, Airoldi G (2007) Atti del seminario Le produzioni di biomassa da coltivazioni arboree a ciclo breve in Piemonte Lombriasco (TO) Italy
5. Bidini G, Cotana F, Buratti C, Fantozzi F, Barbanera M (2006) Analisi del ciclo di vita del pellet da SRF attraverso misure dirette dei consumi energetici 61° Congresso Nazionale ATI—Perugia Italy 12–15 Settembre
6. BIOCOLT, AA.VV. (2010) Colture energetiche per il disinquinamento della laguna di Venezia. Technical report, Veneto Agricoltura
7. Biomass energy report (2009) Politecnico di Milano Dipartimento di Ingegneria gestionale. http://www.energystrategy.it
8. Bonari E, Piccioni E (2006) SRF di pioppo nella pianura litoranea toscana. Principali risultati di alcune esperienze a lungo periodo Sherwood 128:31–36
9. Brun F, Mosso A (2009) Valutazioni economiche della redditività di colture legnose da biomassa e confronto con colture annuali. In: conference La sostenibilità della filiera per la produzione di biomassa da coltivazioni arboree a ciclo breve in Piemonte. Castellamonte (TO) Italy 6 marzo
10. Camcom (2010) Camere di commercio Italiane, Listini prezzi settimanali. http://www.camcom.gov.it
11. Chiaramonti D, Recchia L. (2009) Life cycle analysis of biofuels: can it be the right tool for project assessment? In 17th European Biomass Conference and Exhibition from research to industry and markets, Hamburg, 29 June–2 July 2009
12. Chiaramonti D, Recchia L (2010) Is life cycle assessment (LCA) a suitable method for quantitative CO_2 saving estimations? The impact of field input on the LCA results for a pure vegetable oil chain. Biomass Bioenergy 34:787–797
13. COFEA. AA.VV (2009) La raccolta della biomassa forestale. Tecniche, Economia e Sicurezza del lavoro. Technical report. Monterotondo (RM) Italy, settembre
14. Di Fulvio F (2010) Raccolta del legno: tecnologie per la realtà italiana. MMW 3–4:42–48
15. F.R.I.M.A.T. (2008) federazione regionale imprese meccanizzazione agricola della Toscana–Firenze tariffario delle lavorazioni meccanico-agricole della Toscana
16. Francescato V, Antonini E, Zuccoli Bergomi L (2009) Legna e cippato, Ed. AIEL Associazione italiana nergie agroforestali, Legnano (Pd)
17. Global Emission Model for Integrated Systems (GEMIS) (2010). Oko-Institute.V. http://www.oeko.de/service/gemis/
18. Giorcelli G, Allegro G, Verani S (2008) Aspetti fitosanitari in piantagioni da biomassa. Sherwood 143:11–14

19. Hippoliti G (1997) Appunti di meccanizzazione forestale, Ed. Studio editoriale fiorentino, Firenze
20. L'Informatore Agrario. Supplemento Prezzi e caratteristiche delle macchine operatrici n°05 febbraio 2010
21. L'Informatore Agrario. Supplemento Prezzi e caratteristiche delle macchine operatrici, n° 08 febbraio 2009a
22. L'Informatore Agrario Supplemento Prezzi e caratteristiche delle macchine agricole nuove e usate n° 42 novembre 2009b
23. Magagnotti N, Picchi G, Spinelli R, Lombardini C (2010) Testata di raccolta in prova su pioppo e robinia di 4 anni Supplemento Energia Rinnovabile. L'Informatore Agrario 11:24–26
24. Negrin M, Pettenella D (2010) Produttività, convenienza economica e qualità. Sherwood 163:5–11
25. Pari L, Assirelli A (2009) Pioppo SRF, come ridurre i costi per l'impianto. Agricoltura 6:66–68
26. Pari L, Civitarese V (2009) Falciatrinciacaricatrice Spapperi riveduta e corretta. Supplemento Energia Rinnovabile, L'Informatore Agrario 5:18–21
27. Picchi G, Spinelli R (2010) La logistica su container riduce i costi della biomassa legno. Supplemento Energia rinnovabile, L'Informatore Agrario 17:45–48
28. Picchio R, Baldini S (2009) Utilizzo dei residui forestali a fini energetici, Atti del convegno Interazione fra Selvicoltura & Meccanizzazione Forestale nei paesi del Mediterraneo. Santa Flora (GR) Italy, 26–27 settembre 2002. 181–188
29. P.R.I.O.F. Prezzario regionale per interventi ed opere forestali della Regione Toscana, 2010
30. Spinelli R (2009) C'è posto per la meccanizzazione spinta in Italia? Atti del convegno Interazione fra Selvicoltura & Meccanizzazione Forestale nei paesi del Mediterraneo. Santa Flora (GR) Italy, 26–27 settembre 2002. 153–170
31. Spinelli R, Lombardini C (2009) Tutti i vantaggi nell'esbosco grazie all'uso di un imballatrice. Supplemento Energia rinnovabile a L'Informatore Agrario 27:26–28
32. Spinelli R, Secknus M, Magagnotti N, Hartsough BR, Francescato V, Antonini E, Casini L (2007) Foresta-Legno-Energia Linee guida per lo sviluppo di un modello di utilizzo del cippato forestale a fini energetici. Guidelines book
33. Spinelli R, Nati C, Magagnotti N (2006) SRF di pioppo Macchine e sistemi per la raccolta. Sherwood 128:56–59
34. Spinelli R, Secknus M (2005) Restituire competitività alla biomassa forestale. Alberi e territorio 12:45–49
35. Recchia L, Cini E, Corsi S (2010) Multicriteria analysis to evaluate the energetic reuse of riparian vegetation. Appl Energy 87:310–319
36. Verani S, Sperandio G, Di Matteo G (2010) Analisi del lavoro della Claas Jaguar 880 con testata GBE-1 nella raccolta di un pioppeto da biomassa. Forest 7: 22–27
37. Verani S, Sperandio G (2008) Aspetti tecnico-economici della raccolta delle SRF. In conference Le biomasse agro-forestali: una risorsa sostenibile, Bologna Italy 13 novembre 2008
38. Verani S, Sperandio G (2006) Piantagioni energetiche su piccola scala. Un caso studio nel centro Italia. Sherwood 128:37–41
39. Woodland, AA.VV (2009) La filiera Legno-Energia come strumento di valorizzazione delle biomasse legnose agroforestali, Woodland Energy, Programma Probio–MiPaf

Chapter 5
Olive Oil Production Chain

5.1 Introduction

In Chap. 3, the energetic reuse of agricultural residues coming from olive oil production chain is investigated exploiting both Multicriteria Analysis (MCA) and Life Cycle Assessment (LCA) methodologies. In this Chapter, attention is focused on other aspects of the olive oil working chain: olive grove characteristics, management and productivity; olive harvesting and transport to the extraction plant; olive oil extraction process.

The reference framework for this study is, again, Italy and, in particular, Tuscany. Over the last decades the role of Tuscan agriculture and, consequently that of farms has slowly changed. Originally, farms were only considered as technical-economic units where agricultural, forestry or zootechnical productions were implemented. Nowadays, new tasks have been added to traditional ones: landscape conservation, land coverage and environmental protection by various types of pollution. These tasks are associated with farms not only by national and European policies, but also by market requirements. Actually, Tuscany is characterized by high-quality agricultural products, whose alimentary function is usually joined to the culture of the specific territory in which they are produced. As a consequence, Tuscan agriculture, in particular high-quality products, mostly refers to high-level markets, which are interested in both food product quality and its associated cultural message. For these reasons, recent regional planning policies in agriculture have strongly promoted technological innovations in production processes aimed at both preserving landscape and environment and, at the same time, improving product quality.

Extra-virgin olive oil represents, together with wine, the most typical agricultural product characterizing Tuscany in both national and international markets. Olive grove landscapes and high-quality oil are worldwide associated with the idea of Tuscany. However, in the last decades, economic sustainability of olive oil production chain has become critical, due to several factors: fragmentation of the

L. Recchia et al., *Multicriteria Analysis and LCA Techniques,*
Green Energy and Technology, DOI: 10.1007/978-0-85729-704-4_5,
© Springer-Verlag London Limited 2011

production web, increasing labour cost, difficulties in grove management due to territory peculiarities, increasing national and international competition. Additional efforts are required to apply a philosophy of "total quality", where quality means optimization of all resources required by the production process, which, at the same time, should provide one with essential information for performing strategic planning, as extensively discussed in Chap. 1. As a consequence, developing adequate methodologies to quantify and evaluate environmental sustainability associated with olive oil production chain is essential, considering also the possibility of settling environmental certifications as an additional value to product quality.

In this prospect, recent studies [2–4, 8] have shown how both MCA and LCA methodologies can represent effective tools to analyse olive oil production chain. This can be achieved not without any difficulty, due to the extreme heterogeneity of agricultural processes associated with the variety of environments in which crops grow, to the different typologies of cultivars of each crop, to the various levels of mechanization in field, to the residues management, and so on. In this chapter, an example of the application of these methodologies is provided.

5.2 Multicriteria Analysis to Define and Optimize the Olive Oil Chain

According to the general scheme discussed in Chap. 2, the analysis of the olive oil chain starts from the definition of different possible scenarios concerning the different phases of the chain, that is:

1. Agricultural phase (olive grove typology, fertilization and weed control, olive harvesting and prunings reuse);
2. Olive transport from the grove to the olive oil mill (logistics);
3. Extraction phase (configuration and size of the extraction plant, machine typology, and waste reuse).

Different scenarios can be combined in order to define the whole olive oil production chain. However, as discussed in Chap. 2, the possible scenarios are characterized by a large variability. The application of the MCA methodology is useful to help one in selecting the most significant scenarios according to the following essential requirements which must be set in advance:

- Main goals that must be accomplished;
- Main benefits and drawbacks associated with each scenario;
- A number of evaluation criteria for scenario selection;
- Outcomes and/or scores associated with each scenario for each evaluation criterion.

Scenarios that achieve higher scores represent the most suitable combinations according to the imposed requirements. They are subsequently considered for further evaluation by means of a LCA analysis.

5.2.1 Scenarios Definition

Scenarios for the olive oil production line are defined considering different solutions available for the olive grove, transport from the grove to the mill and the extraction plant. Details of the selection are reported here.

For what concerns the agricultural phase, olive grove configurations are identified taking into account models that can be adapted to the reference territory for which the analysis is performed. In the present application, which refers to the Tuscan territory and agriculture, four different grove models are defined according to Cresti et al. [7]:

1. Marginal olive groves (Fig. 5.1a, b), characterized by severe structural constraints: irregularly shaped trees aged 50 years or more; steeply sloped (over 25%) or terraced grounds located in hill and mountain areas hardly accessible by mechanical devices and requiring high work levels both for maintenance (pruning, scrub control, wall and terrace repairings) and for harvest; highly variable and irregular tree distribution over small areas (usually less than 5 ha);
2. Traditional olive groves (Fig. 5.1c, d), characterized by trees aged between 25 and 50 years distributed over middle-sloped grounds (between 10 and 25%) located in hills and/or rolling plains, easily accessible by most mechanical devices for agricultural operations; typical field size is around 5 ha, with an average tree density of about 250 trees/ha;
3. Intensive olive groves (Fig. 5.2a), characterized by young, regularly shaped plants (aged less than 25 years) distributed over rolling or flat plains sloped less than 10%, where most agricultural operations (both for grove management and for harvest) are mechanized; high tree density (up to 500 trees/ha);
4. Super-intensive olive groves (Fig. 5.2b), which differ from intensive ones mainly for tree density (up to 1,000 trees/ha) and fully mechanized management and harvest; their diffusion, even if still limited in Tuscany, has been increasing in the last few years. Details of the four field scenarios are summarized in Table 5.1.

Collected olives must be carried from the grove to the mill for olive oil extraction within 24 h from harvest, in order to preserve olives from degenerative processes and increase oil quality. As a consequence, olive transport represents a critical step in the whole production chain of high quality extra-virgin oil. Different solutions are possible, depending on the location of olive groves, and the size and typology of olive mills. However, considering the Tuscan situation characterized by small−sized farms spread all over the country, it is common to have olive mills located inside each farm, or at a short distance (usually less than

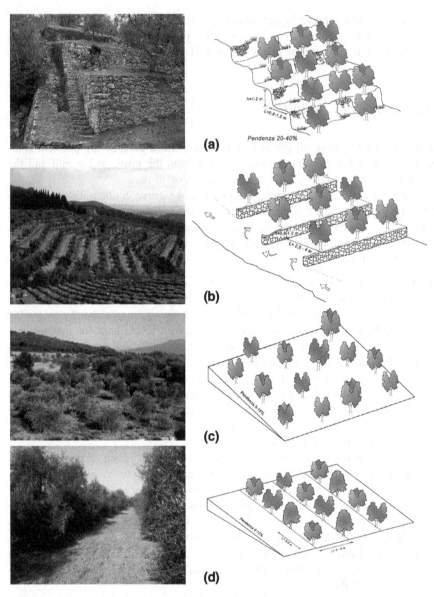

Fig. 5.1 Marginal and traditional olive grove models (from Cresti et al. [7]). **a** Non accessible marginal olive grove (F1 scenario), **b** accessible marginal grove (F1 scenario), **c** irregular traditional grove (F2 scenario), **d** regular traditional grove (F2 scenario)

10 km) from it. Recently, a number of large-sized cooperative mills have been developing, which collect and process olives from wider areas. For these reasons, three different scenarios for olive transport have been identified and considered for the present analysis:

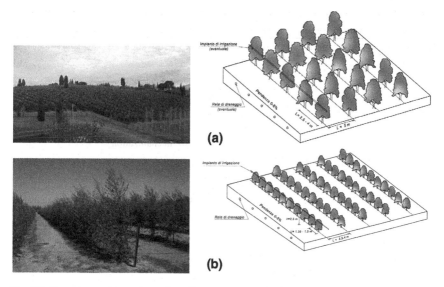

Fig. 5.2 Intensive and superintensive olive grove models (from Cresti et al. [7]). **a** Intensive grove (F3 scenario), **b** superintensive grove (F4 scenario)

1. No transport: olives are processed inside each farm in the local olive mill;
2. Short-distance transport: olives are processed in olive mills located within 10 km from the grove;
3. Medium-distance transport: olives are processed in olive mills at a distance between 10 and 50 km from the grove.

In the first scenario, olives are carried to the mill by the same devices used for field management (mainly farm tractors, Fig. 5.3). Pick-up vans are used in the second scenario, allowed by the short distance between the farm and the mill, and the limited amount of olives typically transported (300–1,000 kg of olives for each lot). For longer distances and larger loads, as those typical of the third scenario, lorries are usually employed. Details of the three transport scenarios are reported in Table 5.2.

Oil extraction is the third phase of the production chain. In the last decades, centrifugal decanters have become widely employed to separate oil from vegetation water and pomace, so the basic layout of modern extraction plants is quite similar for most olive oil mills:

1. Olive defoliation and washing (Fig. 5.4);
2. Olive milling (Fig. 5.5);
3. Kneading (Fig. 5.6);
4. Centrifugal separation (decanter) (Fig. 5.7);
5. Filtration (Fig. 5.8).

However, within this basic layout, different solutions are possible. The use of different milling machines (hammer, stone or disc mills), horizontal or vertical

Table 5.1 Definition of olive grove scenarios (data from Cresti et al. [7] and EFNCP/ARPA [9])

Grove model	Scenario	Field characteristics	Field management and mechanization
Marginal	$F1$	Irregularly shaped trees sometimes in mixed orchards	No fertilization
		Tree age range: $\geq$50 years	No irrigation
		Steeply sloped (>25%)/ terraced grounds with walls	No/occasional pesticide use (copper)
		Typical location: hill and mountain areas	No mechanization
		Typical field size: $\leq$5 ha	No prunings reuse
		High variable tree density	
Traditional	$F2$	Irregularly distributed trees	Organic fertilization (animal manure, leaves, compost, manufactured organic fertilizers)
		Tree age range: 25–50 years	No irrigation
		Middle-sloped (10–25%) grounds	Pesticide use: 2–10 treatments per year
		Typical location: hills and rolling plains	Partly mechanized prunings collection, by–hand harvesting with vibrating poles
		Typical field size: 5–10 ha	No prunings reuse
		Tree density: $\leq$250 trees/ ha	
Intensive	$F3$	Regularly arranged orchards short, with single-stem trees	Mineral fertilization
		Tree age range: $\leq$25 years	Drip-system irrigation
		Little-sloped ($\leq$10%)/flat grounds	Pesticide use: 2–10 treatments per year
		Typical location: rolling/ flat plains	Mechanized prunings collection and harvesting
		Typical field size: $\geq$10 ha	Energetic reuse of prunings
		Tree density: 250–500 trees/ha	
Superintensive	$F4$	Regularly arranged orchards short, with single–stem trees	Mineral fertilization
		Tree age range: $\leq$25 years	Drip–system irrigation with pumping plant
		Little-sloped ($\leq$10%)/flat grounds	Pesticide use: 2–10 treatments per year
		Typical location: flat plains	Fully mechanized agricultural operations
		Typical field size: $\geq$10 ha	Energetic reuse of prunings
		Tree density: 500–1,000 trees/ha and more	

Fig. 5.3 Olives conferred to the mill by means of a farm tractor (*T1* scenario)

Table 5.2 Definition of transport scenarios

Transport	Scenario	Distance	Notes
Local	*T1*	No transport (local mill inside farm)	Olive management operated by farm tractors
Short distance	*T2*	≤10 km	Operated by pick-up vans
Medium distance	*T3*	10–50 km	Operated by lorries

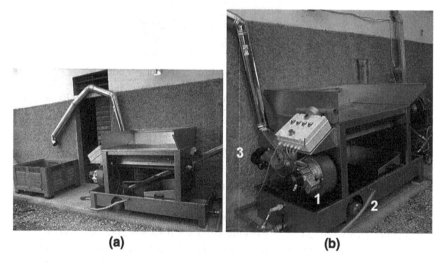

Fig. 5.4 Olive defoliation and washing machine. **a** Overall view, **b** *1* hopper engine for olive feeding; *2* water pump; *3* fan for foliage discharge

kneading, two- or three-phase centrifugal decanters for oil separation, has a significant impact on both mill management and olive oil characteristics. Moreover, waste treatment (vegetation water and pomace) represents another critical point in the extraction phase, especially for what concerns economic and environmental sustainability of the whole process.

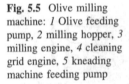

Fig. 5.5 Olive milling machine: *1* Olive feeding pump, *2* milling hopper, *3* milling engine, *4* cleaning grid engine, *5* kneading machine feeding pump

In the present analysis three different mill scenarios are taken into account, referring to typical situations found in Tuscany:

1. A small-sized traditional olive mill, located inside the farm, characterized by a small working capacity (less than ≤ 200 kg$_{olives}$/h), essential technology with a two-phase decanter, no waste reuse with both vegetation water and pomace delivered in field, using natural gas (methane) for both plant heating and sanitary water production (*P*1 scenario);

2. A medium-sized olive mill, which can serve several farms located at a short distance, characterized by a working capacity between 200 and 500 kg$_{olives}$/h with a two-phase decanter, no waste reuse with both vegetation water and pomace delivered in field, but exploiting prunings as biofuel used in the heat boiler (*P*2 scenario);

3. A medium-sized olive mill, similar to the previous one, but equipped with a pomace stone separator (Fig. 5.9), in order to exploit both pomace stone and prunings as biofuels (*P*3 scenario).

The characteristics of each scenario are summarized in Table 5.3. For what concerns *P*3 scenario, thermal energy for both plant heating and sanitary water is supposed to be provided by a wood chips boiler burning prunings only. This

Fig. 5.6 Vertical kneading machines

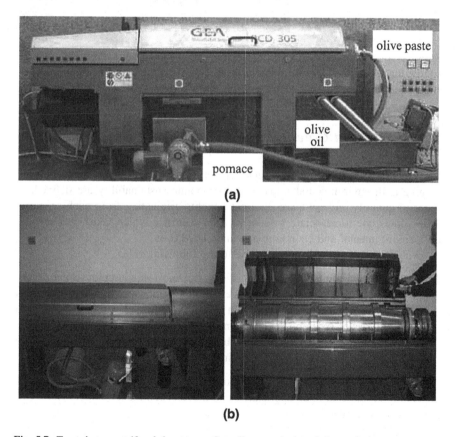

Fig. 5.7 Two-phase centrifugal decanter. **a** Overall external view, **b** internal view

solution allows one to avoid using boilers burning different kinds of biofuels, which usually require a more complex management, with different regulations for each kind of biofuel, have a lower efficiency and a more expensive maintenance.

Fig. 5.8 Paper filter

Stone extracted from pomace can be sold, representing an additional income source for the mill, and, at the same time, having beneficial environmental effects, since it can replace the use of natural gas elsewhere.

5.2.2 Evaluation Criteria Identification

For each scenario illustrated in Sect. 5.2.1 a number of evaluation criteria concerning both environmental impacts and economic sustainability are defined, in order to quantify and compare different benefits originating from each scenario. This allows one to associate a score to each scenario, which takes into account both environmental and economic aspects related to it.

Evaluation criteria are set according to the following rules:

1. From the environmental point of view, the most favourable scenarios are characterized by low mechanization levels of field management, short transport distances and high-efficient extraction plants exploiting energetic reuse of field and plant wastes (prunings and/or pomace stone);
2. For what concerns economic aspects, both field and plant management costs are considered, as well as the whole chain productivity.

Scores associated with each scenario can be determined in different ways, depending on the examined application and the decision maker choice. Here, two methods are considered:

1. By assigning a qualitative score (A, B, C, ...) to different scenarios, from the most to the less favourable one, and combining them by means of a decision–ranking matrix [12];
2. By using numerical scores, the higher one corresponding to the most favourable scenario, and averaging them on the number of criteria for both environmental and economic aspects.

Fig. 5.9 Pomace stone separator. **a** Overview of stone separator, **b** pomace feeding pump, **c** stone output, **d** extracted pomace stone

Three different qualitative scores have been adopted here (*A*, *B*, *C*), corresponding to numerical scores ranging from 3 to 1. Details of adopted evaluation criteria and their corresponding scores are reported in Table 5.4. These criteria have been chosen considering the scenarios taken into account in the present analysis (Tables 5.1, 5.2, 5.3), and exploiting data reported in the literature.

Table 5.3 Definition of extraction plant scenarios

Plant model	Scenario	Plant characteristics
Small-sized	P1	Working capacity: $\leq$200 kg$_{olives}$/h
		Two-phase decanter
		No pomace stone separation
		By-products (pomace and vegetation water) treated as waste
		Heat boiler with natural gas as fuel
Medium-sized	P2	Working capacity: 200–500 kg$_{olives}$/ h
		Two-phase decanter
		No pomace stone separation
		By-products (pomace and vegetation water) treated as waste
		Heat boiler with olive prunings as biofuel
Medium-sized with pomace stone separation	P3	Working capacity: 200–500 kg$_{olives}$/h
		Two-phase decanter
		Pomace stone separation
		Pomace residue and vegetation water treated as waste
		Heat boiler with olive prunings and pomace stone as biofuel

Table 5.4 Definition of evaluation criteria for olive oil chain scenarios

Criteria	Description	Scenario typology	Weights		
			$A = 3$	$B = 2$	$C = 1$
Environmental aspects					
CA1	Irrigation	F	No irrigation	Without pumping plant	With pumping plant
CA2	Fertilization	F	No fertilization	Organic fertilizers	Mineral fertilizers
CA3	Harvest method	F	Manual	Partly mechanized (vibrating poles)	Fully mechanized
CA4	Transport distance	T	No transport (local olive mill)	$\leq$10 km	>10 km
CA5	Waste reuse	P	Prunings and pomace stone	Prunings	No reuse
Economic aspects					
CE1	Olive harvest costs	F	$\leq$50 €/q	50–70 €/q	$\geq$70 €/q
CE2	Prunings management costs	F	$\leq$70 €/t	70–110 €/t	$\geq$110 €/t
CE3	Olive yield	F	$\geq$6 t/ha	4–6 t/ha	$\leq$4 t/ha

Environmental aspects divide traditional, low-mechanized olive groves from intensive, highly mechanized ones, which reuse both field and extraction plant wastes as sources of renewable energy, while reducing the amount of fossil fuels employed in the production process.

For what concerns the economic aspects, numerical values for the selected criteria have been chosen referring to existing databases. In particular, values for each field scenario are determined as follows.

5.2.2.1 Olive Harvest Costs and Olive Yield

Harvest costs depend on several factors: tree distribution and density in the grove, average olive yield for each tree, harvest method (manual or mechanized). Cresti et al. [7] provide a detailed discussion about harvest cost evaluation for different olive grove typologies, reporting harvest costs by hectare of field and by quintal of olives as a function of the harvest method. Here, the following assumptions for each scenario are made:

$F1$ scenario: small field size (≤ 5 ha) with ancient trees characterized by a very large canopy, producing more than 20 kg_{olives}/tree; manual harvest with 10 workers; global harvest cost: 80 $€/q_{olives}$ [7]; maximum olive yield: 1.5 t/ha [9];

$F2$ scenario: field size 5 ha with old trees characterized by a large regular canopy, producing 20 kg_{olives}/tree; tree density 250 trees/ha; partially mechanized harvest (vibrating poles) with 10 workers; global harvest cost: 70 $€/q_{olives}$ [7]; olive yield: 5 t/ha [9];

$F3$ scenario: field size 10 ha with uniformly-distributed trees characterized by a regularly pruned canopy, producing 10 kg_{olives}/tree; tree density 500 trees/ha; fully mechanized harvest (olive shaker and gathering umbrella); global harvest cost: 40 $€/q_{olives}$ [7]; olive yield: 5 t/ha [9];

$F4$ scenario: field size 20 ha with young, uniformly-distributed trees characterized by a regularly pruned canopy, producing 8 kg_{olives}/tree; tree density 800 trees/ha; fully mechanized harvest (olive shaker and gathering umbrella); global harvest cost: 35 $€/q_{olives}$ [7]; olive yield: 6.4 t/ha [9].

5.2.2.2 Prunings Management Costs

Prunings management in olive groves is analysed and discussed in detail in Chap. 3. Here, some data are considered in order to provide a rough estimation of its economic impact on the whole production chain, referring to the examined four field scenarios and considering an average prunings production of about 3 t_{wb}/(ha year):

$F1$ scenario: prunings are collected, piled up and and burned in field by hand, requiring a considerable amount of manpower. Considering a manpower

Table 5.5 Decision ranking matrix adopted for criteria scores combination

	A	B	C
A	A	A	B
B	A	B	C
C	B	C	C

cost of 12 €/h and a productivity of 0.18 t/h both for prunings collection and piling and for prunings burning [7], the estimated overall cost amount exceeds 130 €/t;

F2 scenario: prunings are collected by hand, piled up together by means of a tractor equipped with forks, and manually burned in field. For this case, considering the same values for manpower unitary cost ad productivity as for the F1 scenario, and a cost of about 50 €/h with a productivity of 1 t/h for the tractor (see Cresti et al. [7] and Chap. 3), the estimated overall cost amount is about 100 €/t;

F3 and F4 scenarios: both scenarios assume a fully mechanized prunings management with prunings reused as biofuel. For what concerns management costs, the significant increase in productivity achieved exploiting a tractor equipped with forks for both collection, piling and transport (1 t/h) allows one to abate costs to about 60 €/t [7].

5.2.3 Olive Oil Chains Evaluation

The olive oil production chain can be made up considering different field, transport and plant scenarios for each phase of the chain, selected among those proposed in Sect. 5.2.1. In order to apply MCA to the whole production chain, as described in Chap. 2, it is essential to associate a global score with each chain arrangement, so that all arrangements can be compared to identify which ones have a higher level of sustainability, according to the adopted evaluation criteria. As a consequence, it is essential to define suitable rules to combine scores associated with each scenario in order to obtain the required global one.

If qualitative scores (A, B or C) are used, score combination is usually achieved by means of a decision ranking matrix [12]. The matrix used in the present application is reported in Table 5.5. It is a symmetrical matrix used to combine values two by two, obtaining one final global score for the whole chain. In some cases, it is possible to obtain different final scores by changing the order of couple evaluation, since this is a non-commutative operation. If this happens, the lowest possible value for the global score is chosen for conservative reasons.

Due to the large variability of possible evaluation criteria, only three qualitative scores are adopted in the present application. This choice allows one to combine different weights in a relatively simple way, reducing risks of ambiguous results in the determination of the final score associated with each chain arrangement, as discussed before. However, one of the main drawbacks of this simplified approach

Table 5.6 Olive oil chain evaluation throughout MCA

Chain	Scenarios			Environmental score						Economic score				Global score
				CA1	CA2	CA3	CA4	CA5	Total	CE1	CE2	CE3	Total	
1	F	T1	P1	A	A	A	A	C	**B**	C	C	C	**C**	**C**
				3	3	3	3	1	**2.60**	1	1	1	**1.00**	**1.80**
2	F1	T2	P1	A	A	A	B	B	**B**	C	C	C	**C**	**C**
				3	3	3	2	1	**2.40**	1	1	1	**1.00**	**1.70**
3	F2	T1	P1	A	B	B	A	C	**B**	B	B	B	**B**	**B**
				3	2	2	3	1	**2.20**	2	2	2	**2.00**	**2.10**
4	F2	T1	P2	A	B	B	A	B	**A**	B	B	B	**B**	**A**
				3	2	2	3	2	**2.40**	2	2	2	**2.00**	**2.20**
5	F2	T2	P1	A	B	B	B	C	**B**	B	B	B	**B**	**B**
				3	2	2	2	1	**2.00**	2	2	2	**2.00**	**2.00**
6	F2	T2	P2	A	B	B	B	B	**A**	B	B	B	**B**	**A**
				3	2	2	2	2	**2.20**	2	2	2	**2.00**	**2.10**
7	F3	T2	P2	B	C	C	B	B	**C**	A	A	B	**A**	**B**
				2	1	1	2	2	**1.60**	3	3	2	**2.67**	**2.13**
8	F3	T2	P3	B	C	C	B	A	**B**	A	A	B	**A**	**A**
				2	1	1	2	3	**1.80**	3	3	2	**2.67**	**2.23**
9	F3	T3	P2	B	C	C	C	B	**C**	A	A	B	**A**	**B**
				2	1	1	1	2	**1.40**	3	3	2	**2.67**	**2.03**
10	F3	T3	P3	B	C	C	C	A	**B**	A	A	B	**A**	**A**
				2	1	1	1	3	**1.60**	3	3	2	**2.67**	**2.13**
11	F4	T2	P2	C	C	C	B	B	**C**	A	A	A	**A**	**B**
				1	1	1	2	2	**1.40**	3	3	3	**3.00**	**2.20**
12	F4	T2	P3	C	C	C	B	A	**B**	A	A	A	**A**	**A**
				1	1	1	2	3	**1.60**	3	3	3	**3.00**	**2.30**
13	F4	T3	P2	C	C	C	C	B	**C**	A	A	A	**A**	**B**
				1	1	1	1	2	**1.20**	3	3	3	**3.00**	**2.10**
14	F4	T3	P3	C	C	C	C	A	**B**	A	A	A	**A**	**A**
				1	1	1	1	3	**1.40**	3	3	3	**3.00**	**2.20**

is that most arrangements may have the same score, preventing one from identifying the most significant ones and making the MCA approach ineffective. For this reason, additional numerical scores are associated with each criterion level, as described in Sect. 5.2.2. The final global score is determined both qualitatively, using the decision matrix of Table 5.5, and quantitatively, by averaging scores associated to each criterion level on the number of environmental and economic criteria separately, and then making a final average between the two. Qualitative and quantitative results are eventually compared.

Combining all scenarios identified for each phase of the olive oil chain, 14 different configurations are obtained (Table 5.6). A number of additional configurations have been discarded, since they do not represent realistic cases in applications. For example, all configurations in which the F1 scenario (marginal olive grove) is combined to a medium−sized olive mill with or without stone separator (P2 and P3 scenarios) are not taken into account. On the contrary, intensive and superintensive groves are only coupled with medium-sized mills collecting olives from a wider area than the single farm, possibly providing by-product reuse facilities.

For each configuration reported in Table 5.6, the total scores obtained by applying either environmental (from $CA1$ to $CA5$) or economic (from $CE1$ to $CE3$) criteria described in Table 5.4 are provided, both as a qualitative (A, B or C) and as a quantitative numerical result. A global score is eventually determined by combining the two partial scores by exploiting the decision matrix (Table 5.5) or averaging numerical ones.

The obtained global score is used for comparing configurations all together, and for identifying the most favourable ones, which are selected for the LCA analysis. As discussed in Chap. 3, there is no reason for giving more importance to either environmental or economic scores, since their corresponding criteria are completely independent of each other, and they are not comparable from a technical point of view. The introduction of an additional score to distinguish the contribution of these two factors can be justified only on a political basis, if decision makers decide to place environmental aspects before economic ones, or vice versa.

The results of the MCA analysis identify five configurations as more interesting for olive oil chain sustainability: n.4, 8, 11, 12 and 14. The highest score (A level, 2.30 points) is obtained by configuration n. 12. This is characterized by an excellent result in terms of economic criteria, due to a highly mechanized field management and a high olive yield, and by an average environmental score, mainly due to both prunings and pomace stone reuse. Configuration n. 8 (second in place: A level, 2.23 points) differs from the first one only for grove characteristics (intensive grove, $F3$ scenario), and for the use of a waterfall irrigation system without pumping plant. The remaining three configurations (n. 4, 11 and 14) obtain the same numerical score (2.20 points), even if n. 11 has a slightly lower qualitative score (B level) than the other two. However, while n. 11 and n. 14 are quite similar in terms of both environmental and economic features, configuration n.4 only, among the 5, has a total environmental score higher than the economic one. Its economic drawbacks, due to the low level of mechanization and low olive yields, are balanced by a lower environmental impact due to traditional grove management (absence of irrigation plants, organic fertilization) and simpler logistics, with an olive mill inside the farm which does not require transport and exploits bio-fuel for the heat boiler.

5.3 LCA Methodology Applied to the Olive Oil Chain

The LCA methodology, whose details are described and discussed in Chap. 2, is applied to the five best configurations identified by means of the MCA analysis, i.e. chains n. 4, 8, 11, 12 and 14 shown in Table 5.6. As for the other applications, the main target of LCA is to determine which one of these five chains have a lower impact on the environment, in terms of both $CO_{2\ eq}$ emissions and CER.

The application of LCA is carried out using the software GEMIS 4.5 [11]. This software exploits a large database collecting information about European bio-energy chains and biofuels. In order to use GEMIS, it is essential to set up an inventory phase. This operation consists in the collection of a number of data concerning all processes involved in the selected chains, which must be provided to the GEMIS software as inputs for performing LCA. Inventory data refers to process inputs in terms of raw materials required for field operations (water, fertilizers, pesticides), fuel and energy consumptions (oil, electricity, biomass) and wastes. A normalization of all data is required with respect to the unit product output of each process (i.e. the functional unit) in order to have results of LCA normalized as well.

Collection of data for the inventory phase is often a hard task to accomplish, due to the large variability in process inputs inside the same configuration, and the difficulty of finding consistent and complete data for all processes. At the same time, this operation represents a crucial point in the application of LCA methodology, since lack or inconsistency of data may turn out into incorrect evaluation of both $CO_{2\,eq}$ and CER outputs. For these reasons, particular care should be paid in collecting information for each chain configuration, cross-referencing data from different sources, if possible.

Data concerning the olive oil chain can be divided into four main groups

1. Olive grove data;
2. Prunings reuse data;
3. Transport data;
4. Olive mill data.

For what concerns the second group, prunings reuse, data are provided and discussed in detail in Chap. 3. The other data are presented here.

5.3.1 Olive Grove and Transport Data

Grove data required for performing LCA are:

- Olive yield;
- Water consumption;
- Type (organic/mineral) and amount of employed fertilizers, distinct for active principle;
- Pesticide consumption;
- Fuel consumption.

Three different field scenarios are considered in the five chain configurations selected using MCA: $F2$, $F3$ and $F4$. Data concerning each scenario are reported and discussed here.

For what concerns olive yield, input data are computed for each scenario considering tree density and tree average production, reported in Sect. 5.2.2.

Data are normalized with respect to the annual olive production of 1 ha of olive grove:

$F2$ scenario: considering a tree density of 250 trees/ha and a tree production of 20 kg_{olives}/ (tree year), an overall production of 5,000 kg_{olives}/(ha year) is estimated;

$F3$ scenario: a tree density of 500 trees/ha and an average tree production of 10 kg_{olives}/ (tree year) lead to an overall production of 5,000 kg_{olives}/(ha year), equivalent to $F2$ scenario;

$F4$ scenario: in this case, the higher tree density of about 800 trees/ha compensates the lower average tree production of 8 kg_{olives}/tree year, resulting in a higher overall production of 6,400 kg_{olives}/(ha year).

The results of this evaluation are consistent with the data provided by EFNCP/ ARPA [9], as reported in Sect. 5.2.2.

Water consumption for irrigation is not easy to quantify, since it is strongly variable, depending on many factors, such as the geographical area where the grove is located, its climatic and topographic conditions, average rainfall, evapo-transpiration and temperatures, soil types, water availability along the year, and so on. For the present analysis, two field scenarios exploit grove irrigation ($F3$ and $F4$). A rough estimate of water consumption for the two is obtained considering data provided by EFNCP/ARPA [9], which gives a water amount range between 1,500 m^3/(ha year) and 5,000 m^3/(ha year), relying on several case studies carried out in the Mediterranean area. Considering the minimum and maximum amounts for $F3$ and $F4$ scenarios, respectively, and normalizing data with the annual olive yield per hectare, an estimate of 300 kg_{water}/ kg_{olives} for $F3$ scenario, and 781 kg_{water}/ kg_{olives} for $F4$ scenario is obtained.

For what concerns fertilization, organic fertilizers are employed in $F2$ scenario and mineral fertilizers in both $F3$ and $F4$ scenarios. Data concerning both fertilizers quantities and active principles are obtained by Recchia et al. [13], who collected experimental data in a number of Tuscany farms. Similar quantities and principles are assumed for all scenarios. Normalized fertilizer amounts per kg of produced olives are reported in Table 5.7.

Similar considerations as for fertilizers can be made for the amount of pesticides. In Tuscany, many farms either do not perform plant health control, or exploit basic sanitary treatments using copper sulphate. Here, an average quantity of 6 kg of copper sulphate per hectare is assumed for all scenarios, relying on data available from some Tuscany farms. Normalized amounts per kg of produced olives are provided in Table 5.7.

The evaluation of fuel consumption in field management is more complex, since it must take into account several operations carried out with a variety of devices. In the present application, fuel consumptions related to prunings operations and management, fertilization and disinfestation, and olive harvest are considered for the three scenarios $F2$, $F3$ and $F4$. Inventory data are presented in Tables 5.8, 5.9, 5.10.

The global consumption for each field operation in terms of diesel fuel, typically used in agricultural machines, is computed according to Eq. 5.1:

Table 5.7 Summary of grove inventory data for the 5 most sustainable olive oil chains

Chain			4	8	11	12	14
Grove scenarios	Units	Notes	F2	F3	F4	F4	F4
Olive yield	kg/(ha year)		5,000	5,000	6,400	6,400	6,400
Water consumption	kg/kg_{olives}		0.00	300	781	781	781
Fertilizer type			Organic	Mineral	Mineral	Mineral	Mineral
Fertilizer active principle	kg/kg_{olives}	P	0.00275	0.00275	0.00275	0.00275	0.00275
	kg/kg_{olives}	Ca	0.006143	0.006143	0.006143	0.006143	0.006143
	kg/kg_{olives}	N	0.005143	0.005143	0.005143	0.005143	0.005143
	kg/kg_{olives}	K	0.009107	0.009107	0.009107	0.009107	0.009107
Pesticides	kg/kg_{olives}		0.0012	0.0012	0.000938	0.000938	0.000938
Fuel consumption	kWh/kg_{olives}		0.14	0.50	0.55	0.55	0.55

Table 5.8 Fuel consumption data for $F2$ field scenario

Field operation	Notes	Nominal device power (kW)	Average utilization (%)	Usage time (h/ha)	Energy consumption (kWh/ha)
Prunings	Operation management	4.50	60	10.00	64.04
		50.00	60	3.00	213.48
Fertilization		50.00	60	2.50	177.90
Disinfestation		50.00	60	2.50	177.90
Harvest	Vibrating poles	1.25	60	50.00	88.95
Total					722.27

Table 5.9 Fuel consumption data for $F3$ field scenario

Field operation	Notes	Nominal device power (kW)	Average utilization (%)	Usage time (h/ha)	Energy consumption (kWh/ha)
Prunings	Operation management	4.50	60	10.00	64.04
		50.00	60	3.00	213.48
Fertilization		50.00	60	2.50	177.90
Disinfestation		50.00	60	2.50	177.90
Harvest	Shaker with umbrella	60.00	60	21.88	1,867.95
Total					2,501.27

$$\text{Energy consumption} = \text{Nominal device power}(\text{NDP})$$
$$\times \text{ Average utilization}(\text{AU}) \times \text{Specific fuel consumption}(\text{SFP})$$
$$\times \text{ Lower heating value}(\text{LHV}) \times \text{Usage time}(\text{UT}). \qquad (5.1)$$

Table 5.10 Fuel consumption data for $F4$ field scenario

Field operation	Notes	Nominal device power (kW)	Average utilization (%)	Usage time (h/ha)	Energy consumption (kWh/ha)
Prunings	Operation management	10.00	60	2.00	28.46
		30.00	60	3.00	128.09
Fertilization		50.00	60	2.50	177.90
Disinfestation		50.00	60	2.50	177.90
Harvest	Shaker with umbrella	60.00	60	35.00	2,988.72
Total					3,501.07

Reference values for $SFP = 0.2$ kg/kWh and $LHV = 11.86$ kWh/kg are available from the literature [5, 10], and are assumed constant for each device in each scenario, as well as an average device power utilization $AU = 60\%$. The other data are provided by Cresti et al. [7], which exploit data referring to typical Tuscan farms.

Detailing each scenario, $F2$ values reported in Table 5.8 are obtained in the following way. For prunings operation, the use of pneumatic scissors is assumed (Fig. 5.10), with $NDP = 4.5$ kW, considering a usage time $UT = 10$ h/ha (2 h/ha for five workers), while prunings management is performed using a 50 kW farm tractor requiring 3 h/ha. Both fertilization and disinfestation use the same 50 kW tractor capable of treating 100 trees/h for a density of 250 trees/ha. Harvest is managed by means of vibrating poles, ($NDP = 0.250$ kW) for five workers, each employing about 10 h/ha (Fig. 5.11).

$F3$ values (Table 5.9) differ from $F2$ ones only for what concerns harvest management, which is fully mechanized and exploits a 60 kW olive shaker equipped with a gathering umbrella for collecting olives (Fig. 5.12). The value for $UT = 21.88$ h/ha is computed dividing the tree density 500 trees/ha by the machine working capacity (approximately 160 trees/day for a 7 h/day working time).

Considering $F4$ scenario, differences with respect to the other two concern both prunings operations and management, and harvest. Prunings operations are performed by means of a machine equipped with cutting bars with serrated plates ($NDP = 10$ kW) requiring a $UT = 2$ h/ha (Fig. 5.13). Prunings are treated for energetic reuse by exploiting an industrial shredder, capable of collecting, shredding and transporting prunings ($NDP = 30$ kW, $UT = 3$ h/ha) (Fig. 5.14). Olive harvest is achieved by means of an olive shaker with a gathering umbrella, as in $F3$ scenario, but for the required average usage time ($UT = 35.00$ h/ha) due to the higher value of tree density (800 trees/ha).

For transport scenarios $T1$, $T2$ and $T3$, LCA inventory data only concerns employed means of transport and distances from the grove to the olive mill. These data are reported in Tables 5.2 and 5.11. While $T1$ does not require any transport, exploiting farm tractors (Fig. 5.3), $T2$ considers a 5 km distance covered by means of pick-up vans, and $T3$ a 30 km distance with transport operated by lorries.

Fig. 5.10 Pneumatic scissors for prunings operations (from Cresti et al. [7]). **a** Fixed scissors with telescopic arm, **b** pneumatic scissors

(a)

(b)

5.3.2 Olive Oil Mill Data

Olive mill data required by LCA methodology are:

- Oil extraction efficiency;
- Oil production;

Fig. 5.11 Vibrating poles for harvesting operations (from Cresti et al. [7]). **a** Different commercial typologies, **b** vibrating pole usage

- Operative and life time of olive mill device equipment;
- Electricity consumption and type of power plant;
- Heat consumption and type of heat plant;
- Water consumption;
- Waste typology and quantity.

Two extraction plant scenarios are considered in the five chain configurations selected using MCA, $P2$ and $P3$, which essentially differ about pomace stone reuse as biofuel. Data concerning each scenario are reported and discussed here (Table 5.12).

Olive oil extraction efficiency typically varies in a range between 10 and 20% of olive mass flow processed in the mill. Variations are due to several factors (type of cultivar, ripening level, extraction plant management, ...), which are difficult to predict. In this application, an average value of 15% is assumed for both scenarios, which represents a typical value for Tuscan production.

Both olive mills are supposed to be operative for about 3 months a year, from October to December, during olive harvest. In this short period, working times often extend over many hours a day, in order to face peak-time demand from farms. In recent years, this has become a fundamental requirement for extraction plants, since milling olives within 24 h from harvest has been proved to be essential for obtaining high-quality extra virgin olive oil. As a consequence, a

Fig. 5.12 Olive shaker
equipped with a gathering
umbrella (from Cresti et al.
[7])

Fig. 5.13 Farm tractor
equipped with cutting bars
with serrated plates (from
Cresti et al. [7])

daily working time of 16 h a day is considered in the present application, for
30 days a month, resulting in an estimated global amount of working hours
equivalent to 1,440 h/year for both scenarios. Assuming a mill working capacity of
350 kg_{olive}/h and a 15% extraction efficiency, the overall oil production is
75,600 kg_{oil}/year.

Fig. 5.14 Devices for collecting, shredding and transporting prunings (from Cresti et al. [7])

The lifetime of mill equipment varies from device to device. Considering only mechanical devices (olive crusher, kneading machine, centrifugal decanter, pomace stone separator and auxiliary equipments), a lifetime of 4 years is considered a reasonable estimate, according to manufacturers' handbooks and instructions.

For what concerns electricity consumptions, both $P2$ and $P3$ scenarios are supposed to exploit electric grid power. Similar configurations for both plants are considered, with an olive crusher, a vertical kneading machine and a two-phase centrifugal decanter, the only difference being the use of a centrifugal pomace stone separator in $P3$ scenario. Since the working capacity is the same (300 kg$_{olives}$/h), all nominal powers are assumed to be the same for devices present in both plants. Detailed data for two extraction plants similar to $P2$ and $P3$ are provided by Cini et al. [6] and shown in Table 5.13 for completeness. For each operation and sub-operation of the extraction process the nominal device power and its usage percentage and time are reported. Times have been directly measured during extraction operations at the olive oil mill, while usage percentages have been either measured or afterward deduced from global electricity consumptions. Energy consumptions

Table 5.11 Summary of transport plant inventory data for the 5 most sustainable olive oil chains

Chain			4	8	11	12	14
Transport scenarios	Units	Notes	T1	T2	T2	T2	T3
Means of transport			–	Pick-up Van	Pick-up Van	Pick-up Van	Lorry
Distance	km		–	5	5	5	30

are reported both as absolute and specific values per each kilogram of extracted olive oil. As a result, this study estimates an overall average consumption of 0.195 kWh/kg_{oil} and 0.314 kWh/kg_{oil} for P2 and P3 scenarios, respectively. Therefore, the increase in energy consumptions due to the use of a stone separator (P3 scenario) is about 61% of the whole consumption without it (P2 scenario).

Heat consumptions required for mill management are estimated considering an average specific consumption of 50 kWh/ (m^2 year) for both scenarios. This value can be compared to typical values for a residential building in Italy, which are about 200–250 kWh/ (m^2 year). The much lower value is essentially due to the fact that heating of olive mill premises is limited to what is strictly necessary. Assuming a mill surface of 200 m^2, the resulting heat consumption is 0.132 kWh/ kg_{oil}. For what concerns the type of heat plant, P2 scenario is equipped with a biomass heat boiler using prunings. This is also the case of the P3 scenario, where, in addition, stone extracted from pomace is sold as a biofuel, as discussed in Sect. 5.2.1.

Mill water requirements are strongly variable, depending on both olive type and quality, and mill configuration. Focusing the attention on the extraction process, olive washing and oil separation in the centrifugal decanter are the two most water-consuming operations [1]. In the present application, a two-phase centrifugal decanter is considered for oil extraction, which greatly reduces water consumptions with respect to three-phase decanters. Water added at a two-phase decanter input is typically 10% of the pomace mass flow (10 kg of water per 100 kg of processed olives). In addition, 20 kg of water is required for washing and 17 kg for rinsing 100 kg of olives. No water is added in the pomace stone separator (P3 scenario), since water content in the pomace exiting the two-phase decanter is sufficient to perform stone separation. As a result, an overall average specific water consumption of 3.13 kg/kg_{oil} is obtained with 15% extraction efficiency.

Wastes at the end of the extraction process basically consist of vegetation water and pomace. Considering a two-phase decanter processing 100 kg of olives with the addition of 10 kg of water, as discussed before, about 4 kg of vegetation water (0.26 kg/kg_{oil}) and 91 kg of pomace (6.07 kg/kg_{oil}) are typically obtained at the decanter outlet. So, the global waste quantity in P2 scenario sums up to 6.33 kg/kg_{oil}. In P3 scenario, pomace is destoned by means of a centrifugal separator, which is reused as a biofuel and is not treated as waste. Considering an average solid content of about 40% on olive mass, the typical efficiency of the stone extraction process is

Table 5.12 Summary of extraction plant inventory data for the 5 most sustainable olive oil chains

Chain			4	8	11	12	14
Plant scenarios	Units	Notes	P2	P3	P2	P3	P3
Extraction efficiency	%		15	15	15	15	15
Operative time	h/year		1,440	1,440	1,440	1,440	1,440
Life time	year		4	4	4	4	4
Olive oil production	kg/year		75,600	75,600	75,600	75,600	75,600
Electricity power plant			Grid	Grid	Grid	Grid	Grid
Electricity consumption	kWh/kg$_{oil}$		0.195	0.314	0.195	0.314	0.314
Heat power plant		Biomass heat boiler	Prunings	Prunings (pomace stone sold as biofuel)	Prunings	Prunings (pomace stone sold as biofuel)	Prunings (pomace stone sold as biofuel)
Heat consumption	kWh/kg$_{oil}$		0.132	0.132	0.132	0.132	0.132
Water consumption	kg/kg$_{oil}$	2-phase decanter	3.13	3.13	3.13	3.13	3.13
Waste type			Vegetation water and pomace	Vegetation water and destoned pomace	Vegetation water and pomace	Vegetation water and destoned pomace	Vegetation water and destoned pomace
Waste quantity	kg/kg$_{oil}$		6.33	5.42	6.33	5.42	5.42

about 15% on olive mass as well [6]. Consequently, the total amount of wastes decreases to 5.42 kg/kg_{oil}.

5.4 LCA Results

The results of LCA methodology applied to the 5 chains selected by means of MCA are provided and discussed here. In Table 5.13, numerical outputs from the GEMIS software are reported in terms of both $CO_{2\,eq}$ emissions and CER for all chains. The results are also illustrated in Figs. 5.15 and 5.16.

For what concerns $CO_{2\,eq}$ emissions, contributions due to different operations associated with each phase of the production chain (field operations, transport and extraction process) are provided. As a result of LCA analysis, most $CO_{2\,eq}$ emissions are due to field operations. A much lower impact is due to both extraction process and transport, the latter being almost negligible.The total emission amounts are computed by summing up all contributions, which are considered as "debits" in terms of environmental impact. Moreover, for three of the selected chains (n. 8, 12 and 14) a number of "credits" are introduced, related to the reuse of pomace stone as a biofuel in the scenarios composing each chain. Actually, by-product reuse lowers the global amount of $CO_{2\,eq}$ emissions, since it allows one to partially replace the use of non-renewable resources for energy production (oil, natural gas) with renewable ones. On the other hand, prunings are employed in the extraction plant for thermal energy production, and their reuse does not provide any credit, since it is exploited inside the production chain. The total emission amounts accounting for these credits are reported in Table 5.14 and Fig. 5.15.

Comparing results for the five chains, Chain 4 turns out to be the most favourable in terms of both $CO_{2\,eq}$ emissions and CER. As discussed in Sect. 5.2.3 while commenting on MCA results, this chain is the only one having a total environmental score higher than the economic one, due to a low level of mechanization in the grove management. In this chain, the $F2$ scenario does not collect olive tree prunings, which can be either burned or shredded and landfilled. In the first case, emissions associated with burning operations are not considered. In the second case, no increase in soil fertility is taken into account, since the highest contribution is due to organic matter, while no significant amount of nutrients is delivered in the short period to the soil. However, since $P2$ scenario, included in chain 4 as well, makes use of prunings as a biofuel, an external supplying source of prunings is assumed, located at a transport distance lower than 40 km. Related impacts, computed considering the agro-energetic chain n. 18 described in Chap. 3, amount to 38.267 kg $CO_{2\,eq}$ /kWh ($CO_{2\,eq}$ emissions) and 0.077 kWh/kWh(CER).

In Chain 8, which is second on the list, some benefits in $CO_{2\,eq}$ emissions are accounted for considering the reuse of prunings in the extraction phase for thermal energy production ($F3$ scenario). Their evaluation is performed referring again to the agro-energetic chain n. 18 described in Chap. 3.

Table 5.13 Working process powers and energy consumptions for *P2* and *P3* scenarios (data from Cini et al. [6])

Operation	Sub-operation	Nominal device power (kW)	Average utilization (%)	Usage time (h)	Energy consumption	
					Absolute (kWh)	Specific (kWh/ kg_{oil})
Defoliation-Washing		1.77	100	0.36	0.63	0.0140
Crushing	Cutting	5.50	51	0.36	1.00	0.0224
	Auxiliary equipment	1.86	100	0.36	0.66	0.0147
Kneading		1.10	100	1.00	0.63	0.0245
Extraction	Centrifugation	5.00	56	0.54	4.49	0.1003
	Auxiliary equipment	1.50	100	0.54	0.80	0.0179
Filtration		0.75	100	0.07	0.05	0.0011
P2 Total		**27.5**	**66.3**	**2.31**	**8.74**	**0.195**
Stone separation		15.00	66.3	0.54	5.32	0.1186
P3 Total		**42.5**	**66.3**	**2.85**	**14.06**	**0.314**

Fig. 5.15 CO_2 $_{eq}$ emissions for the five most suitable olive oil chains

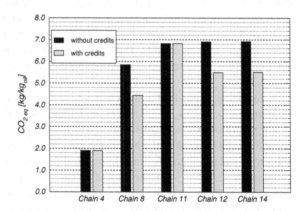

Chains 11, 12 and 14 essentially differ for the extraction plant scenario (*P2* for Chain 11, *P3* for Chains 12 and 14). For what concerns wastes produced during olive oil extraction, they are not contemplated in the present analysis, since spreading in field is provided for in both scenarios, and no differences can be highlighted between them. In fact, the same quantity of nutrients are delivered to soil, since pomace stone contributions are negligible. The chain results in terms of CO_2 $_{eq}$ emissions are quite similar for all the three, if credits are not taken into account. However, significant credits are obtained for Chains 12 and 14 from the reuse of pomace stone (*P3* scenario), which is sold as a biofuel, as previously

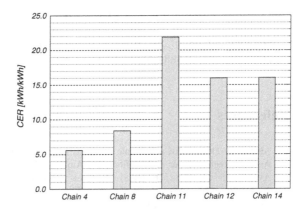

Fig. 5.16 CER emissions for the five most suitable olive oil chains

Table 5.14 CO_2 eq and CER emission results from LCA applied to the five most suitable olive oil chains

	Chain 4	Chain 8	Chain 11	Chain 12	Chain 14
CO_2 eq (kg/kg$_{oil}$)					
Field operations	1.771	5.630	6.686	6.686	6.686
Transport	0.000	0.003	0.003	0.003	0.019
Extraction process	0.134	0.212	0.134	0.212	0.212
Total without credits	1.906	5.846	6.823	6.901	6.917
Total with credits	1.906	4.441	6.823	5.496	5.512
CER (kWh/kg$_{oil}$)	5.589	8.396	21.878	15.957	16.014

discussed in this section and in Sect. 5.2.1. Credits coming from the use of pomace stone instead of natural gas are computed as follows [11]. The amount of CH_4 replaced by pomace stone is given by Eq. 5.2:

$$m_{CH_4} \text{ (kg)} = m_{stone} \text{ (kg)} \times \left(\frac{LHV_{stone}}{LHV_{CH_4}}\right) \times \left(\frac{0.85}{0.95}\right), \quad (5.2)$$

where:

$$LHV_{stone} = 4.5 \text{ kWh/kg,}$$

and

$$LHV_{CH_4} = 13.5 \text{ kWh/kg.}$$

Credits are eventually computed according to GEMIS process *"gas-boiler-CZ-small (2000)"* [11] for both CO_2 eq (Eq. 5.3):

$$Credits_{CH_4}(CO_{2eq}) = -0.30701 \text{ kg } CO_{2eq}/kWh_{CH_4}; \quad (5.3)$$

and CER (Eq. 5.4):

$$\text{Credits}_{CH_4}(\text{CER}) = -0.077 \text{ kWh/kWh}_{CH_4}. \tag{5.4}$$

The total amount of credits is -1.405 kg CO_2 $_{eq}$/kg$_{oil}$, which is subtracted from the total emission results of chains 8, 12 and 14 in Table 5.14.

References

1. ARPAL (2006) Analisi ambientale di comparto produttivo. L'olio d'oliva. Technical report, Agenzia Regionale per la Protezione dell'Ambiente Ligure (ARPAL)
2. Avraamides M, Fatta D (2008) Resource consumption and emissions from olive oil production: a life cycle inventory case study in cyprus. J Cleaner Prod 16:809–821
3. Chiaramonti D, Recchia L (2009) Life cycle analysis of biofuels: can it be the right tool for project assessment? In: 17th European biomass conference and exhibition from research to industry and markets, Hamburg, 29 June–2 July 2009
4. Chiaramonti D, Recchia L (2010) Is life cycle assessment (LCA) a suitable method for quantitative CO_2 saving estimations? The impact of field input on the LCA results for a pure vegetable oil chain. Biomass Bioenergy 34:787–797
5. Cini E, Recchia L (2008) Energia da biomassa: un'opportunità per le aziende agricole
6. Cini E, Recchia L, Daou M, Boncinelli P (2008) Human health benefits and energy savings in olive oil mills. In: International conference on "Innovation technology to empower safety, health and welfare in agriculture and agro-food systems", Ragusa, Italy, September 15–17
7. Cresti G, Gucci R, Zorini LO, Polidori R, Vieri M, Sarri D, Rimediotti M (2009) Progetto MATEO. Modelli tecnici ed economici per la riduzione dei costi di produzione nelle realtà olivicole della Toscana. Technical report, ARSIA Toscana
8. Dessane D (2003) Energy efficiency and life cycle analysis of organic and conventional olive groves in the Messara Valley, Crete, Greece. Master's thesis, Wageningen University
9. EFNCP/ARPA (2000) The environmental impact of olive oil production in the EU: practical options for improving the environmental impact. Technical report, European Forum on Nature Conservation and Pastoralism and the Asociación para el Análisis y Reforma de la Política Agro-rural
10. ENAMA (2005) Prontuario dei consumi di carburante per l'impiego agevolato in agricoltura. Technical report, Ente Nazionale per la Meccanizzazione Agricola
11. GEMIS (2010) Global emission model for integrated systems (GEMIS). Oko-Institut e.V. http://www.oeko.de/service/gemis/
12. Recchia L, Cini E, Corsi S (2010) Multicriteria analysis to evaluate the energetic reuse of riparian vegetation. Appl Energy 87:310–319
13. Recchia L, Gaertner S, Corsi S (2010) LCA applied to agro-industrial products: the case of the vegetable oils in Tuscany. In: LCAfood 2010—VII International conference on life cycle assessment in the agri-food sector, Bari, Italy, September 22–24

Chapter 6
Oil Palm Farming Chain

6.1 Introduction

This chapter focuses on an Multicriteria analysis (MCA) based decision scheme and the related life cycle assessment (LCA), aimed to compare some possible scenarios that planners could have to face when evaluating the convenience of different options when establishing an oil palm plantation.

Higher yields and lower production costs, if compared to other edible oil sources, have focused great interest in oil palm cultivation since the 1970s [10], mainly in Southeast Asia Countries, but also in its areas of origin of Western and Central Africa and in more recently exploited Latin America and Caribbean. In the last years the increase in demand for vegetable oil for biodiesel and the support given by Governments of the main producing Countries, donors, funding institutions and the UN to the large-scale agro-industrial model, have given even more impulse to its expansion making it the fastest growing monocrop plantation in the tropics, with almost 15 Mha standing in 2009 [2].

Due to the uncontrolled booming of commercial plantations that has taken place at the end of the last century and to the consequent negative environmental impact, in the last decades large projects for oil palm have often been criticized and opposed [9]. The need for a more careful and responsible approach, that takes into account not only economical results but also environmental and social aspects of planting in large scale, is nowadays universally recognized and new investments cannot ignore these aspects. This positive response to the so-called sustainable approach and its rules has reduced the land soundly suited for new plantations and has called for more careful analysis of cultivation inputs and techniques, making it necessary, in some cases, to take into account even the need for irrigation.

Large-scale plantations are now also facing a controversial moment where traditional employment of manual labour for most field operations is confronted with increasing labour costs, local unavailability of sufficient labour force and

L. Recchia et al., *Multicriteria Analysis and LCA Techniques*,
Green Energy and Technology, DOI: 10.1007/978-0-85729-704-4_6,
© Springer-Verlag London Limited 2011

technological development that lead to consider mechanization more convenient than before and in some cases absolutely necessary.

The above-mentioned issues are just some out of many that make the establishment of an oil palm plantation a complex exercise which requires evaluating the convenience of different options and alternatives.

The following study wants to show, in a simplified, didactical way, how to confront, with an MCA and LCA approach, a number of possible variables that planners need to take into account when making decisions. It is not aimed to present a real case study or to provide "ready to use" analysis packages, so when applying the proposed method to a real case, even if considering similar scenarios, their description must be thoroughly detailed and relevant data verified according to the specific situation.

6.2 An MCA Application for Evaluating Different Options in Planning an Oil Palm Plantation

6.2.1 Context and Assumptions

This example shows an MCA application for evaluating the convenience of establishing an oil palm plantation in an area where conditions are not ideal and different solutions have to be considered.

A case that is quite realistic at present and is linked to the above−mentioned fact that the extensive planting that has taken place in the last decades, together with the environmental limitations, make land ideally suited for oil palm scarce in some parts of the world, is that often also non ideal conditions have to be taken into account.

One important aspect could concern the amount of rainfall, which could be insufficient for optimal plant growth and production with the consequent need to take into account the possibility of irrigation, while another constraint could be the non availability of contiguous plots of land that does not allow to establish a unique plantation, but obliges to split the planned surface into different locations.

As previously mentioned, also the possibility of mechanizing some routine operations, traditionally carried out by hand, such as weed control and fertilizing, is an option that at the present cannot be neglected.

Concerning the setting up of an oil mill that normally is associated to a plantation, especially in those areas where oil palm is not already a widely established crop, the options are many and complex. In this example only a few are considered, aiming to provide an example of choices touching economic, environmental, energetic and logistics aspects, such as the convenience of providing the mill with a kernel processing facility, of treating the effluents before releasing them in the fields, of burning the empty fruit bunches left after the separation of the fruits, to supply steam and electricity, and of building a centralized plant compared to several smaller installments in the case of fragmented layout of plots.

Aiming to focus on the topics highlighted above, the reference plantation hypothesized for the analysis consists in two plots of about 2,500 ha, 50 km away from each other.

6.2.2 Scenarios

The first step in the MCA approach concerns the definition of various and alternative scenarios based on the different possibilities that must be evaluated.

In our case, following the main options that have been described, the issues to be considered at the first approach would be the convenience of mechanizing field operations, the possibility of providing water through irrigation, the management of logistics and the definition of oil mill size and functions.

6.2.2.1 Field Operations

In an established plantation the main operations consist in fertilizing, weeding, collection and transportation of fruit bunches from the field to the mill [1, 4].

The simplest technique, where operations are mostly manual, is based on spreading by hand the fertilizer around the palms, slashing the weeds with cutlasses or spraying them with herbicides with the use of knapsack sprayers, and transporting the manually harvested fruit bunches to the collection point on the road with the use of a wheelbarrow or other muscle-powered means of transport. The fruit bunches are then collected from the road and transported to the mill by trucks of different type and size.

On the other side, a mechanized management of these operations would be based on the spreading of fertilizers with the use of a tractor-driven broadcaster, the control of weeds through a mower or a mulcher in the case of mechanical control and by the use of a tractor trailed sprayer in the case of chemical control, while the transport to the road would be done with the use of motorized carriers. Harvesting and transport to the mill would not differ from those previously described since manual harvesting is still very difficult to replace and at present there is no convenient alternative to trucks (or agricultural tractor) for road transport for the plantation layout that has been considered.

Considering these two extreme options together with some intermediate possibilities, four possible scenarios have been designed for field operations, as listed in Table 6.1.

Among various possibilities the characteristics proposed for each scenario are:
M1 Weed control is done by mulching with an agricultural tractor with a horizontal axle flail mower.
Spreading of fertilizer is done by an agricultural tractor with a broadcaster. The spreading equipment is followed by another tractor with a trailer carrying the fertilizer.
Collection of fruit bunches and transport to the main road is done by motortrailers.

Table 6.1 Scenarios for field operations

Scenario	Description	Equipment
M1	Fully mechanized	Tractor with mulcher
		Tractor with broadcaster
		Tractor with trailer
		Motortrailers
		Transport of bins to mill
M2	Partially mechanized 1	Tractor with sprayer and guns
		Tractor with broadcaster
		Tractor with trailer
		Transport of bins to mill
M3	Partially mechanized 2	Tractor with mulcher
		Tractor with broadcaster
		Tractor with trailer
		Transport of bins to mill
M4	Not mechanized	Slashers with cutlasses
		Backpack sprayers
		Transport of bins to mill

Transport to mill is done by tractors or trucks that continuously shuttle from the main service roads of the plantation to the mill.

M2 Weed control is done chemically with the use of a tractor trailing a sprayer with two hand guns connected for localized spraying.

Spreading of fertilizer as for M1.

Collection of fruit bunches and transport to the main road is done manually.

Transport to mill as for M1.

M3 Weed control as for M1.

Spreading of fertilizer as for M1.

Collection of fruit bunches and transport to the main road as for M2.

Transport to mill as for M1.

M4 Weed control is manual by slashing and spraying.

Spreading of fertilizer is manual by collecting fertilizers in baskets and wheelbarrows from a parked trailer or bin.

Collection of fruit bunches and transport to the main road is as for M2.

Transport to mill as for M1.

Of course classification of scenarios is only indicative of the main source of work, as it is not possible to completely apply or eliminate the use of motorized equipment (i.e. manual loading of motortrailers or transportation by tractor and trailer of the fertilizer for manual spreading).

6.2.2.2 Water Management

In the hypothesis of planting where rainfall is not sufficient for best expression of oil palm potentiality, production is expected to be lower than optimum, which implies considering the possibility of providing extra water through irrigation.

Table 6.2 Scenarios for water management

Scenario	Description
W1	Rain fed
W2	Permanent irrigation
W3	Emergency irrigation

Table 6.3 Scenarios for logistics

Scenario	Description	Mill number and size	Carrier	Distance (km)
L1	Local	Two small	Tractors	3
L2	Distant	One medium	Trucks	53

Irrigation practiced on such a large scale has high costs and has to be confronted with the expected increase in production. An almost intermediate possibility is that of irrigating only when conditions become critical for the plantation and allow some water deficit for the remaining time.

Following these considerations three scenarios have been considered as listed in Table 6.2.

The different scenarios are characterized by the need for an irrigation system and the usage of water:

W1   No irrigations scheme exists and production is fully dependant on climate course.

W2   Existence of an irrigation system and supply of water to fulfil requirements for optimal growth.

W3   Existence of an irrigation system and sporadic irrigation in the driest periods to avoid water stress.

6.2.2.3 Logistics

The fragmentation of the plantation in two different areas, with a distance of about 50 km, makes it necessary to choose between installing one small mill in each location or one bigger centralized mill in one of the two places.

Many considerations can be made on the convenience of either solution, among them the more complex logistics required by one centralized mill and the higher cost of a double installation.

The first aspect calls for two scenarios: two short range fleets for transport within a short distance or one mixed fleet for local and medium range transport with different implications on costs and organization.

Table 6.3 summarizes the options taken into account for logistics.

From the logistic point of view the two scenarios differ for the type of carrier and the average distance to be covered for reaching the mill.

L1   Agricultural tractors tow trailers or bins to the mill covering an average distance of about 3 km within the plantation (two way trip).

L2   Road truck carry swap-body trailers from the plantation without mill to the one where the mill is located, while agricultural tractors tow trailers or bins within this plantation. In the first case the distance is 103 km (two way trip), in the second case it is 3 km with a resulting average of 53 km.

As for other information calculations have been simplified[1] the purpose of this example being the application of the methodology and not the estimation of a cost or of any other value.

6.2.2.4 Oil Mill

In the mill oil palm fruits are treated with steam and separated from the bunches that are discarded. Fruit processing proceeds with pressure application for juice extraction from fruits and separation of expressed fibre from kernels. The oil is recovered from the juice by dynamic and static decantation; fibre is normally used for burning in the boiler for steam and electricity production while kernels can be processed in a specific plant for recovery of palm kernel oil or sold as a by-product. The waste water from the process is discarded [1].

As mentioned, process variables are many and this is the reason why an MCA approach can help to tackle the complexity of the choice. In this case three possible options have been considered, in addition to the previously introduced alternative between having one or two installations, for defining the scenarios; out of the 16 possible combinations[2] five have been chosen as more interesting for the scope.

Table 6.4 lists these scenarios and their main characteristics.

The number and size of plants has been explained in the previous paragraph, while the presence or not of PKO processing unit does not need explanations. The other two options that have been considered are explained below.

T1–T4 Effluent treatment is a process that is usually applied in bigger plants, but sometimes neglected in smaller ones. Liquid waste contains minerals and organic matter that if correctly treated in specific lagoons can be used for biogas production and for fertilizing the soil, allowing recovering part of the nutrients removed with harvest; on the other side these substances can pollute the soil and water if directly applied in high concentrations.

T2 Use of empty fruit bunches for burning with the purpose of producing steam and energy is usually not practiced because the fibre is normally sufficient for the needs of the mill, but in some cases, where the mill provides electric energy to nearby structures (i.e. villages, other installations, etc.), the availability of an extra source of combustible could be very useful containing the need for fossil fuel (Figs. 6.1–6.5).

---

[1] A 2,500 ha plot can be considered as a square of 5,000 m × 5,000 m with the mill located at the centre where an average two way trip can be estimated in about 3 km. In the case of the transport from one plantation to the other 100 km have to be considered and added to the 3 km of average transport distance inside the origin plantation.

[2] In order to contain the complexity of the exercise the cases where out of the two installations only one could be equipped with the various add-in processes, has not been considered.

Table 6.4 Scenarios for oil mill

Scenario	Description
T1	One medium size plant
	PKO processing
	Effluent treatment
	No EFB energetic conversion
T2	One medium size plant
	PKO processing
	Effluent treatment
	EFB energetic conversion
T3	One medium size plant
	No PKO processing
	Effluent treatment
	No EFB energetic conversion
T4	Two small-sized plants
	No PKO processing
	Effluent treatment
	No EFB energetic conversion
T5	Two small-sized plants
	No PKO processing
	No effluent treatment
	No EFB energetic conversion

Fig. 6.1 Recently planted Oil palm plot

6.2.3 Definition of Evaluation Criteria

Once the possible options have been defined and organized in a number of different scenarios for each one of the considered aspects (M, W, L and T), one or more keys for evaluation must be chosen and scenarios compared according to them.

Fig. 6.2 Manual harvesting of FFB

Fig. 6.3 Motortrailer unloading FFBs in a bin (swap-body)

Fig. 6.4 Palm oil mill

Also in this case, as in other chapters of this book, the most interesting keys of evaluation appear to be environmental impact and economic sustainability.

For each of these keys some criteria have to be defined and values (numerical or boolean) attributed to allow to compare the different scenarios.

Environmental aspects

Energy inputs	have been considered to have a negative impact; use of field equipment or of pumps for irrigation implies higher energy inputs.
Use of chemicals	as for energy inputs, the use of chemicals has been considered negative from an environmental point of view.
Use of water	its negative impact is proportional to the amount used.
Distance	also the distance covered for transporting the oil palm fruits to the mill has been considered negative since it implies a more intense use of machines.
PKO processing	its positive aspects concern the reduction of mass to be transported out of the mill (PKO instead of whole kernels) and the availability of shells for production of energy locally, but the negative impact of transporting the kernels to another processing facility has not been considered significant.
Effluent treatment	this option is positive because it reduces the pollutant components of the waste and, if the biogas produced during the digestion process is collected, no additional CO_2 is released in the atmosphere.
Use of EFB for energy	positive because where more energy is needed it allows to reduce the use of fossil fuel and there is no need for transporting the large mass of empty bunches to the field, but there are some drawbacks, since there is no organic matter returning to the field and burning of bunches together with fibre might not be so efficient, this is why the two options have not been considered fully positive or fully negative.

Economic aspects

Field operations costs	in this example full mechanization has been considered to reduce overall costs because of the lower need for labour and the possibility of carrying out operations in a faster and more productive way. Partially mechanized and manual operations have been considered to have

	similar costs with mulching more expensive than spraying because of the need for much higher power.[3]
Income increase	irrigation allows for higher production, proportionally to the amount of water supplied (total amount require-ments or only deficit reduction), but costs rise at the same time so the balance is not necessarily positive. A low income increase has not been considered favourable because of the high investments required and the risks.[4]
Distance	costs are higher for transporting to a distant facility but transport by truck is cheaper than transport by tractors, so the differences are not considered to be so significant.
Number of plants	two small plants have much higher building and running costs than one medium one.
PKO processing	it is assumed that processing of PKO has a positive balance, but selling the kernels still provides some profit.
Effluent treatment	negative since it implies the setting up and running of complex treatment plants.
Use of EFB for energy	positive because of the value of energy and because the mineral content of the bunches can be cheaply replaced by chemical fertilizers.

Concerning the values, since the operations that have been proposed are not widespread in oil palm cultivation and reliable data appears not to be accessible, numerical values have not been used but scenarios ranked according to expected results. Once more these results are only indicative and their purpose is to give an example of application of the methodology and not to provide an assessment on oil palm cultivation inputs.

Table 6.5 shows the different scenarios classified by expedience (A = most favourable, D = less favourable) according to each criteria. To each class a numeric value (1–4) must be assigned to allow successive calculations and comparison of chains (see Table 6.8).

---

[3] The following figures show the costs (euro/year) hypothesized for each scenario:

Scenario	Mechanization	Labour	Total
M1	663,000	864,000	1,527,000
M2	412,000	1,224,000	1,636,000
M3	435,000	1,224,000	1,659,000
M4	164,000	1,440,000	1,604,000

[4] The following figures show costs and yield hypothesized for each scenario:

Scenario	Cost of water	Production
W1	N/A	$14 \text{ t ha}^{-1} \text{ year}^{-1}$
W2	0.09 € m^3	$18 \text{ t ha}^{-1} \text{ year}^{-1}$
W3	0.11 € m^3	$16 \text{ t ha}^{-1} \text{ year}^{-1}$

Table 6.5 Expedience of scenarios according to evaluation criteria

Criteria	Classification			
	A = 1	B = 2	C = 3	D = 4
Environmental aspects				
Field operations: energy inputs	Low: M4	–	–	High: M1, M2, M3
Field operations: use of chemicals	No: M1, M3	–	–	Yes: M2, M4
Water management: energy inputs	None: W1	–	Medium: W3	High: W2
Water management: usage of water	None: W1	–	Medium: W3	High: W2
Logistics: distance	Short: L1	–	–	Long: L2
Oil mill: PKO processing	Yes: T1, T2	–	No: T3, T4, T5	–
Oil mill: effluent treatment	Yes: T1, T2, T3, T4	–	–	No: T5
Oil mill: EFB for energy	–	Yes: T2	No: T1, T3, T4	–
Economic aspects				
Field operations: costs	Low: M1	–	Medium: M2, M4	High: M3
Water management: income increase	N/A: W1	–	Low: W2	Negative: W3
Logistics: transport costs	–	Lower: L1	Higher: L2	–
Oil mill: no. of plants	One: T1, T2, T3	–	–	Two : T4, T5
Oil mill: PKO processing	Yes: T1, T2	–	No: T3, T4, T5	–
Oil mill: effluent treatment	No: T5	–	–	Yes: T1, T2, T3, T4
Oil mill: EFB for energy	Yes: T2	–	–	No: T1, T3, T4

Table 6.6 Shows an example of attribution of weights to the criteria used in this exercise

Criteria and indicators	Weights
Environmental aspects	
M Energy inputs	0.10
M Use of chemicals	0.10
W Energy inputs	0.10
W Use of water	0.20
L Distance	0.10
T PKO processing	0.05
T Effluent treatment	0.20
T EFB for energy	0.15
Total	1.00
Economical Aspects	
M Yearly cost	0.10
W yearly income increase	0.10
L Costs	0.15
T No. of plants	0.20
T PKO processing	0.10
T Effluent treatment	0.15
T EFB for energy	0.20
Total	1.00

Again this paragraph provides an example of the framework that is followed in MCA while values given are only indicative in order to provide an example of application.

6.2.4 Attribution of Weights

Different scenarios can now be compared on the basis of the criteria that have been proposed and of their evaluation, but it is evident that not all criteria have the same weight in determining the convenience of a scenario. For example, from an environmental point of view, the presence of an effluent treatment plant would be more positive than the installation of a PKO processing facility.

For this reason the contribution of each criterion in determining the overall favourability of a scenario should be considered. Also in this case attribution is subjective and depends on the experience of analysts and on the scope of the analysis.

Table 6.6 shows an example of attribution of weights to the criteria used in this exercise.

6.2.5 Results

Out of the four aspects and the 14 scenarios that have been proposed, 120 layouts of the plantation (chains) are possible. Some of these are contradictory and cannot be considered.

Table 6.7 possible chains and incompatibilities (grey)

chain	Scenarios			
no.	M	W	L	T
1	M1	W1	L1	T1
2	M1	W1	L1	T2
3	M1	W1	L1	T3
4	M1	W1	L1	T4
5	M1	W1	L1	T5
6	M1	W1	L2	T1
7	M1	W1	L2	T2
8	M1	W1	L2	T3
9	M1	W1	L2	T4
10	M1	W1	L2	T5
11	M1	W2	L1	T1
12	M1	W2	L1	T2
13	M1	W2	L1	T3
14	M1	W2	L1	T4
15	M1	W2	L1	T5
16	M1	W2	L2	T1
17	M1	W2	L2	T2
18	M1	W2	L2	T3
19	M1	W2	L2	T4
20	M1	W2	L2	T5
21	M1	W3	L1	T1
22	M1	W3	L1	T2
23	M1	W3	L1	T3
24	M1	W3	L1	T4
25	M1	W3	L1	T5
26	M1	W3	L2	T1
27	M1	W3	L2	T2
28	M1	W3	L2	T3
29	M1	W3	L2	T4
30	M1	W3	L2	T5
31	M2	W1	L1	T1
32	M2	W1	L1	T2
33	M2	W1	L1	T3
34	M2	W1	L1	T4
35	M2	W1	L1	T5
36	M2	W1	L2	T1
37	M2	W1	L2	T2
38	M2	W1	L2	T3
39	M2	W1	L2	T4
40	M2	W1	L2	T5
41	M2	W2	L1	T1
42	M2	W2	L1	T2
43	M2	W2	L1	T3
44	M2	W2	L1	T4

(continued)

Table 6.7 (continued)

chain	Scenarios			
no.	M	W	L	T
45	M2	W2	L1	T5
46	M2	W2	L2	T1
47	M2	W2	L2	T2
48	M2	W2	L2	T3
49	M2	W2	L2	T4
50	M2	W2	L2	T5
51	M2	W3	L1	T1
52	M2	W3	L1	T2
53	M2	W3	L1	T3
54	M2	W3	L1	T4
55	M2	W3	L1	T5
56	M2	W3	L2	T1
57	M2	W3	L2	T2
58	M2	W3	L2	T3
59	M2	W3	L2	T4
60	M2	W3	L2	T5
61	M3	W1	L1	T1
62	M3	W1	L1	T2
63	M3	W1	L1	T3
64	M3	W1	L1	T4
65	M3	W1	L1	T5
66	M3	W1	L2	T1
67	M3	W1	L2	T2
68	M3	W1	L2	T3
69	M3	W1	L2	T4
70	M3	W1	L2	T5
71	M3	W2	L1	T1
72	M3	W2	L1	T2
73	M3	W2	L1	T3
74	M3	W2	L1	T4
75	M3	W2	L1	T5
76	M3	W2	L2	T1
77	M3	W2	L2	T2
78	M3	W2	L2	T3
79	M3	W2	L2	T4
80	M3	W2	L2	T5
81	M3	W3	L1	T1
82	M3	W3	L1	T2
83	M3	W3	L1	T3
84	M3	W3	L1	T4
85	M3	W3	L1	T5
86	M3	W3	L2	T1
87	M3	W3	L2	T2
88	M3	W3	L2	T3

(continued)

Table 6.7 (continued)

chain	Scenarios			
no.	M	W	L	T
89	M3	W3	L2	T4
90	M3	W3	L2	T5
91	M4	W1	L1	T1
92	M4	W1	L1	T2
93	M4	W1	L1	T3
94	M4	W1	L1	T4
95	M4	W1	L1	T5
96	M4	W1	L2	T1
97	M4	W1	L2	T2
98	M4	W1	L2	T3
99	M4	W1	L2	T4
100	M4	W1	L2	T5
101	M4	W2	L1	T1
102	M4	W2	L1	T2
103	M4	W2	L1	T3
104	M4	W2	L1	T4
105	M4	W2	L1	T5
106	M4	W2	L2	T1
107	M4	W2	L2	T2
108	M4	W2	L2	T3
109	M4	W2	L2	T4
110	M4	W2	L2	T5
111	M4	W3	L1	T1
112	M4	W3	L1	T2
113	M4	W3	L1	T3
114	M4	W3	L1	T4
115	M4	W3	L1	T5
116	M4	W3	L2	T1
117	M4	W3	L2	T2
118	M4	W3	L2	T3
119	M4	W3	L2	T4
120	M4	W3	L2	T5

Table 6.7 shows all possible layouts highlighting the incompatible ones.

In Table 6.8 values have been assigned to the criteria of each compatible chain and averages and totals calculated. The first section shows the average of non-weighted values calculated on the basis of values attributed to each criteria, the second section shows the result of values attributed to criteria multiplied for their weight. This is the most important data and is used to identify the more convenient chains (lower values). As it can be noted by comparing the ranking of the chains in the two sections, the best seven are the same, but in a different order, showing how attribution of weight is very important in determining a chain's expediency

Table 6.8 Average (without weights) and total (with weights) values for compatible chains. Best results for weighted calculation are highlighted together with correspondent non-weighted average values

Chain no.	Scenarios				Without weights											With Weights		
					Environmental criteria					Economic criteria					Total	Env. value	Econ. value	Total value
	M	W	L	T	M	W	L	T	Avg val	M	W	L	T	Avg val	Avg val			
4	M1	W1	L1	T4	4	1	1	3	1,88	1	1	2	4	2,71	2,29	1,70	3,00	2,35
5	M1	W1	L1	T5	4	1	1	3	2,25	1	1	2	4	2,29	2,27	2,30	2,55	2,43
6	M1	W1	L2	T1	4	1	4	1	2,00	1	1	3	1	2,14	2,07	1,90	2,35	2,13
7	M1	W1	L2	T2	4	1	4	1	1,88	1	1	3	1	1,71	1,79	1,75	1,75	1,75
8	M1	W1	L2	T3	4	1	4	3	2,25	1	1	3	4	2,43	2,34	2,00	2,55	2,28
14	M1	W2	L1	T4	4	4	1	3	2,63	1	3	2	4	3,00	2,81	2,60	3,20	2,90
15	M1	W2	L1	T5	4	4	1	3	3,00	1	3	2	4	2,57	2,79	3,20	2,75	2,98
16	M1	W2	L2	T1	4	4	4	1	2,75	1	3	3	1	2,43	2,59	2,80	2,55	2,68
17	M1	W2	L2	T2	4	4	4	1	2,63	1	3	3	1	2,00	2,31	2,65	1,95	2,30
18	M1	W2	L2	T3	4	4	4	3	3,00	1	3	3	4	2,71	2,86	2,90	2,75	2,83
24	M1	W3	L1	T4	4	3	1	3	2,38	1	4	2	4	3,14	2,76	2,30	3,30	2,80
25	M1	W3	L1	T5	4	3	1	3	2,75	1	4	2	4	2,71	2,73	2,90	2,85	2,88
26	M1	W3	L2	T1	4	3	4	1	2,50	1	4	3	1	2,57	2,54	2,50	2,65	2,58
27	M1	W3	L2	T2	4	3	4	1	2,38	1	4	3	1	2,14	2,26	2,35	2,05	2,20
28	M1	W3	L2	T3	4	3	4	3	2,75	1	4	3	4	2,86	2,80	2,60	2,85	2,73
34	M2	W1	L1	T4	4	1	1	3	2,25	3	1	2	4	3,00	2,63	2,00	3,20	2,60
35	M2	W1	L1	T5	4	1	1	3	2,63	3	1	2	4	2,57	2,60	2,60	2,75	2,68
36	M2	W1	L2	T1	4	1	4	1	2,38	3	1	3	1	2,43	2,40	2,20	2,55	2,38
37	M2	W1	L2	T2	4	1	4	1	2,25	3	1	3	1	2,00	2,13	2,05	1,95	2,00
38	M2	W1	L2	T3	4	1	4	3	2,63	3	1	3	4	2,71	2,67	2,30	2,75	2,53
44	M2	W2	L1	T4	4	4	1	3	3,00	3	3	2	4	3,29	3,14	2,90	3,40	3,15
45	M2	W2	L1	T5	4	4	1	3	3,38	3	3	2	4	2,86	3,12	3,50	2,95	3,23

(Continued)

Table 6.8 (continued)

Column groups — Chain no. | Scenarios (M, W, L, T) | Without weights [Environmental criteria (M, W, L, T, Avg val); Economic criteria (M, W, L, T, Avg val); Total Avg val] | With Weights (Env. value, Econ. value, Total value)

no.	M	W	L	T	M	W	L	T		Avg val	M	W	L	T		Avg val	Total Avg val	Env. value	Econ. value	Total value
46	M2	W2	L2	T1	4	4	4	1	3	3,13	3	3	3	1	1	2,71	2,92	3,10	2,75	2,93
47	M2	W2	L2	T2	4	4	4	1	2	3,00	3	3	3	1	1	2,29	2,64	2,95	2,15	2,55
48	M2	W2	L2	T3	4	4	4	3	3	3,38	3	3	3	1	3	3,00	3,19	3,20	2,95	3,08
54	M2	W3	L1	T4	4	3	1	1	3	2,75	3	4	2	4	3	3,43	3,09	2,60	3,50	3,05
55	M2	W3	L1	T5	4	3	1	3	3	3,13	3	4	2	4	3	3,00	3,06	3,20	3,05	3,13
56	M2	W3	L2	T1	4	3	4	1	3	2,88	3	4	3	1	1	2,86	2,87	2,80	2,85	2,83
57	M2	W3	L2	T2	4	3	4	1	2	2,75	3	4	3	1	1	2,43	2,59	2,65	2,25	2,45
58	M2	W3	L2	T3	4	3	4	3	3	3,13	3	4	3	1	3	3,14	3,13	2,90	3,05	2,98
64	M3	W1	L1	T4	4	1	1	1	3	1,88	4	1	2	4	3	3,14	2,51	1,70	3,30	2,50
65	M3	W1	L1	T5	4	1	1	3	3	2,25	4	1	2	4	3	2,71	2,48	2,30	2,85	2,58
66	M3	W1	L2	T1	4	1	1	1	3	2,00	4	1	3	1	1	2,57	2,29	1,90	2,65	2,28
67	M3	W1	L2	T2	4	1	1	1	2	1,88	4	1	3	1	1	2,14	2,01	1,75	2,05	1,90
68	M3	W1	L2	T3	4	1	1	3	3	2,25	4	1	3	1	3	2,86	2,55	2,00	2,85	2,43
74	M3	W2	L1	T4	4	4	4	1	3	2,63	4	3	2	4	3	3,43	3,03	2,60	3,50	3,05
75	M3	W2	L1	T5	4	4	4	3	3	3,00	4	3	2	4	3	3,00	3,00	3,20	3,05	3,13
76	M3	W2	L2	T1	4	4	4	1	3	2,75	4	3	3	1	1	2,86	2,80	2,80	2,85	2,83
77	M3	W2	L2	T2	4	4	4	1	2	2,63	4	3	3	1	1	2,43	2,53	2,65	2,25	2,45
78	M3	W2	L2	T3	4	4	4	3	3	3,00	4	3	3	1	3	3,14	3,07	2,90	3,05	2,98
84	M3	W3	L1	T4	4	3	1	1	3	2,38	4	4	2	4	3	3,57	2,97	2,30	3,60	2,95
85	M3	W3	L1	T5	4	3	1	3	3	2,75	4	4	2	4	3	3,14	2,95	2,90	3,15	3,03
86	M3	W3	L2	T1	4	3	4	1	3	2,50	4	4	3	1	1	3,00	2,75	2,50	2,95	2,73
87	M3	W3	L2	T2	4	3	4	1	2	2,38	4	4	3	1	1	2,57	2,47	2,35	2,35	2,35
88	M3	W3	L2	T3	4	3	4	3	3	2,75	4	4	3	1	3	3,29	3,02	2,60	3,15	2,88

(Continued)

Table 6.8 (continued)

Chain no.	Scenarios				Without weights											Total Avg val	With Weights		
	M	W	L	T	Environmental criteria				Avg val	Economic criteria				Avg val			Env. value	Econ. value	Total value
					M	W	L	T		M	W	L	T						
94	M4	W1	L1	T4	1	4	1	3	1,88	3	1	2	4	3,00	2,44	1,70	3,20	2,45
95	M4	W1	L1	T5	1	4	1	3	2,25	3	1	2	4	2,57	2,41	2,30	2,75	2,53
96	M4	W1	L2	T1	1	4	4	1	2,00	3	1	3	1	2,43	2,21	1,90	2,55	2,23
97	M4	W1	L2	T2	1	4	4	1	1,88	3	1	3	1	2,00	1,94	1,75	1,95	1,85
98	M4	W1	L2	T3	1	4	4	3	2,25	3	1	3	4	2,71	2,48	2,00	2,75	2,38
104	M4	W2	L1	T4	1	4	1	3	2,63	3	3	2	4	3,29	2,96	2,60	3,40	3,00
105	M4	W2	L1	T5	1	4	1	3	3,00	3	3	2	4	2,86	2,93	3,20	2,95	3,08
106	M4	W2	L2	T1	1	4	4	1	2,75	3	3	3	1	2,71	2,73	2,80	2,75	2,78
107	M4	W2	L2	T2	1	4	4	1	2,63	3	3	3	1	2,29	2,46	2,65	2,15	2,40
108	M4	W2	L2	T3	1	4	4	3	3,00	3	3	3	4	3,00	3,00	2,90	2,95	2,93
114	M4	W3	L1	T4	1	3	1	3	2,38	3	4	2	4	3,43	2,90	2,30	3,50	2,90
115	M4	W3	L1	T5	1	3	1	3	2,75	3	4	2	4	3,00	2,88	2,90	3,05	2,98
116	M4	W3	L2	T1	1	3	4	1	2,50	3	4	3	1	2,86	2,68	2,50	2,85	2,68
117	M4	W3	L2	T2	1	3	4	1	2,38	3	4	3	1	2,43	2,40	2,35	2,25	2,30
118	M4	W3	L2	T3	1	3	4	3	2,75	3	4	3	4	3,14	2,95	2,60	3,05	2,83

Table 6.9 Inventory of I/O for the processes

Input/output	Unit	Chains						
		7	97	67	37	6	27	96
Field operations								
Yield	tFFB/ha	14	14	14	14	14	16	14
EFB as fertiliser	Yes/No	No	No	No	No	Yes	No	Yes
Water consumption	kg/kgFFB	0	0	0	0	0	15.75	0
N	kg/kgFFB	0.00500	0.00500	0.00500	0.00500	0.00312	0.00500	0.00312
P	kg/kgFFB	0.00060	0.00060	0.00060	0.00060	0.00058	0.00060	0.00058
K	kg/kgFFB	0.00700	0.00700	0.00700	0.00700	0.00700	0.00700	0.00700
Mg	kg/kgFFB	0.00135	0.00135	0.00135	0.00135	0.00135	0.00135	0.00135
Herbicides/pesticides	kg/kgFFB	0.00013	0.00013	0.00013	0.00013	0.00013	0.00013	0.00013
Fuel consumption	l/kgFFB	0.00263	0.00130	0.00173	0.00163	0.00263	0.00230	0.00130
Oil mill processing								
Water	kg/kgCPO	4	4	4	4	4	4	4
Total electricity requirement	kWh/kgCPO	781	781	781	781	781	684	781
Electricity from power central	kWh/kgCPO	773	773	773	773	773	676	773
Electricity from the grid	kWh/kgCPO	9	9	9	9	9	8	9
Steam from power central	kWh/kgCPO	14,680	14,680	14,680	14,680	14,680	12,845	14,680
Diesel for motors	kWh/kgCPO	134	134	134	134	134	103	134
POME treatment								
Electricity	kWh/kgPOME	0.1050	0.1050	0.1050	0.1050	0.1050	0.1050	0.1050
Power central								
Diesel fuel	kWh/kWh	0.0922	0.0922	0.0922	0.0922	0.0922	0.0922	0.0922
Water	kg/kWh	15.40	15.40	15.40	15.40	15.40	15.40	15.40
Palm kernel oil processing								
Electricity from the grid	kWh/kgPKO	0.2672	0.2672	0.2672	0.2672	0.2672	0.2672	0.2672

Table 6.10 Nutrients in EFB that can be returned to the soil (d.b. = dry basis, w.b. = wet basis) [5]

	Production (%)	N amount (kg/tEFB)	P amount (kg/tEFB)
EFB	22.0% of FFB	8.00 d.b.	0.96 d.b.
		3.20 w.b. (60%)	0.38 w.b. (60%)

6.3 Life Cycle Assessment

The life Cycle Assessment methodology can provide some important information about $CO_{2\,eq}$ emissions and CER originated as a consequence of the inputs implied in each scenario. This allows to analyse more in depth the environmental impact of the selected combination, especially the global impact which is less evident but equally important as the local one.

6.3.1 Inventory

The first step of applying LCA to the more convenient combinations resulting from the MCA, is the definition of an inventory, that is to say a list of process inputs and outputs in terms of the raw materials required for field operations (water, fertilizers, pesticides), fuel and energy consumptions (oil, electricity, biomass) and wastes produced.

Once inputs and outputs have been defined, reliable values must be associated with each voice. This can be done on the basis of the analyst's experience or with the use of the data collected in process measuring or in bibliography, although most of this information can be found in specific databases associated with LCA tools which provide data for many crops and processes.

In this case, as for previous applications in this book, Gemis 4.5 [3] has been used and the main inventory data are shown in Table 6.9 according to the processes implied in the scenarios that have been taken into account in the MCA.

All the inputs are based on the literature review [5–8] and on Gemis 4.5 database, but all data can be modified according to the analyst's aims or to any specific data that he holds.

Particularly, when the EFB are used as fertilisers, a reduction of the mineral nutrients has been hypothesised according to the contributions estimated for the EFB to the soil shown in Table 6.10.

6.3.2 Results

Tables 6.11 and 6.12 and Figs. 6.6 and 6.7 show the results of LCA application.

Table 6.11 CO$_2$ equivalent emissions with and without credits for the selected chains

Chain	Combination	CO$_2$eq (g/kg_oil)		CO$_2$eq (g/kWh_oil)						
		Total with credits	Total without credits	Total with credits	Total without credits	Field phase	Transport	Extraction phase	POME treatment	Kernel Milling
6	M1-W1-L2-T1	–	**521.01**	–	**50.49**	20.64	1.29	26.17	0.02	2.37
7	M1-W1-L2-T2	**−37.25**	593.09	**−3.61**	57.48	27.56	1.29	26.24	0.02	2.37
27	M1-W3-L2-T2	**−87.31**	557.60	**−8.46**	54.00	27.08	1.29	23.24	0.02	2.37
37	M2-W1-L2-T2	**−52.25**	578.08	**−5.06**	56.03	26.12	1.29	26.24	0.02	2.37
67	M3-W1-L2-T2	**−50.89**	579.45	**−4.93**	56.17	26.25	1.29	26.24	0.02	2.37
96	M4-W1-L2-T1	–	**500.97**	–	**48.55**	18.71	1.29	26.17	0.02	2.37
97	M4-W1-L2-T2	**−57.29**	573.05	**−5.55**	55.55	25.63	1.29	26.24	0.02	2.37

Table 6.12 CER with and without credits for the selected chains

Chain	Combination	CER (Wh/kg_oil)		CER (Wh/kW_oil)	
		Total with credits	Total without credits	Total with credits	Total without credits
6	M1-W1-L2-T1	–	**1,748.43**	–	**169.44**
7	M1-W1-L2-T2	**−394.79**	1,884.84	**−38.26**	182.66
27	M1-W3-L2-T2	**−576.29**	1,756.07	**−55.85**	170.18
37	M2-W1-L2-T2	**−450.20**	1,829.43	**−43.63**	177.29
67	M3-W1-L2-T2	**−445.14**	1,834.49	**−43.14**	177.78
96	M4-W1-L2-T1	–	**1,674.44**	–	**162.27**
97	M4-W1-L2-T2	**−468.77**	1,810.86	**−45.43**	175.49

Fig. 6.5 Heap of EFB in the mill's yard

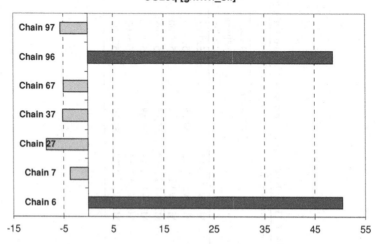

Fig. 6.6 CO_2 equivalent emissions for the selected chains

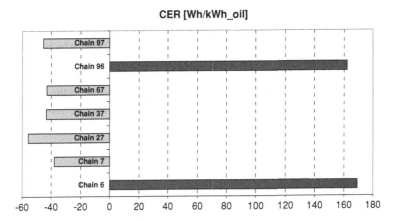

Fig. 6.7 CER equivalent emissions for the selected chains

Table 6.10 shows a prediction of the amount of CO_2 generated by each single chain referred to the oil produced and to its energy value. CO_2 emissions are well known to be a negative indicator so the best chain under this point of view is no. 96 and the main reason is to impute to the absence of mechanization; however, this chain is the worst among the seven selected in the previous paragraph because of the negative impact of manual work from an economic point of view.

For the same reason the higher emissions are connected to full field mechanization as in chain no. 7 that was selected as the most convenient in MCA.

Things change radically when carbon credits are considered: the worst chains become no. 6 and 96 while the best is no. 27. In the first case the reason is that no credits are assigned when EFB is redistributed in the field, although a reduction in the need for mineral fertilisers has been considered as previously noted (Tables 6.9, 6.10).

Figure 6.6 shows in a very evident way the difference between the chains where EFB are not burned (T1) and the others (T2).

From Table 6.11 it can also be seen that many entries are common to all chains since L2 is the only option represented while T1 and T2 only differ for the reutilization of the EFB that has some influence on the overall extraction process. Also the field phase is influenced by the use made of the EFB as can be seen by comparing chains no. 6 and no. 7 which only differ for this operation. Concerning the water management it can be seen how production affects the CO_2 balance: as a matter of fact the increase in productivity of irrigated fields evens up and exceeds the higher emissions caused by the irrigation system and its operation, as is evident by comparing chains 7 and 27.

Since LCA is principally aimed to evaluate a process from an environmental point of view, it is interesting to note that chains 7, 67 and 97 which emerged as the most environmental friendly in the MCA, are instead the worst ones in terms of

CO_2 emission. This is not necessarily to ascribe to a mistake in attribution of values and weights (see Tables 6.5, 6.6) but to as explained in Sect. 2.7.

Table 6.12 and Fig. 6.7 show, in a similar way, the CER, which is the total fossil energy needed for the operation or phase, and its relation with CO_2 emissions can be considered by comparing them with the previous tables.

Finally, it must be highlighted that all the obtained results of the LCA do not consider any contribution due to the site characteristics, e.g. carbon stocks variations caused by the land use change.

References

1. Corley RHV, Tinker PB (2007) The oil palm. Blackwell, Oxford
2. FAO (2010) Faostat Database. http://www.faostat.fao.org
3. GEMIS (2010) http://www.oeko.de/service/gemis
4. Rankine I, Fairhurst T (1998) IPNI oil palm series field handbook vol.3: Mature. PPI/PPIC, Singapore and Agrisoft, Australia
5. Reinhardt G, Rettenmaier N, Gärtner S (2007) Rain Forest for Biodiesel? IFEU, Germany
6. Schmidt JH (2007) Life cycle assessment of rapeseed oil and palm oil. Ph.D. thesis. Department of Development and Planning, Aalborg University, Denmark
7. Singh G, Huan LK, Leng T, Kow DL (1999). Oil palm and the environment: a Malaysian perspective, Malaysian Oil Palm Growers Council, Kuala Lumpur
8. Sumathi S, Chai SP, Mohame AR (2008) Utilization of oil palm as a source of renewable energy in Malaysia. Renew Sustain Energy Rev 12(2008):2404–2421
9. Tauli-Corpuz V, Tamang P (2007) Oil palm and other commercial biofuels and commodity markets tree plantations, monocropping: impacts on indigenous peoples' land tenure and resource management systems and livelihoods. UN Permanent Forum on Indigenous Issues, Sixth session. New York, USA. http://www.un.org/esa/socdev/unpfii/documents/6session_crp6.doc
10. Thoenes P (2006) Palm oil focus. FAO, Commodities and Trade Division. Rome, Italy

Index

L. Recchia et al., *Multicriteria Analysis and LCA Techniques,*
Green Energy and Technology, DOI: 10.1007/978-0-85729-704-4,
© Springer-Verlag London Limited 2011